NASA SP-4220

Wingless Flight
The Lifting Body Story

R. Dale Reed

with Darlene Lister

Foreword by General Chuck Yeager

The NASA History Series

National Aeronautics and Space Administration
NASA History Office
Office of Policy and Plans
Washington, DC 1997

Library of Congress cataloging-in-Publication Data

Reed, R. Dale, 1930
Wingless Flight: The Lifting Body Story/R. Dale Reed,
with Darlene Lister: foreword by Chuck Yeager.
p. cm.—(SP; 4220. NASA History Series)

Includes bibliographical references (p.) and index.
1. Lifting bodies—United States—Design and construction—
History. 2. NASA Dryden Flight Research Center—History.
I. Lister, Darlene, 1945- . II. Title. III. Series: NASA SP; 4220.
IV. Series: NASA Historical Series.
TL713.7.R44 1997 629.133'3—dc21

For sale by the U.S. Government Printing Office
Superintendent of Documents, Mail Stop: SSOP, Washington, DC 20402-9328
ISBN 0-16-049390-0

DEDICATED TO BRAVERY AND FAITH

Paul Bikle

A leader who believed in our cause and put his career at high risk by having faith in us to meet our commitments.

Milt Thompson

A research test pilot who not only put his life at risk but exhibited complete faith in wingless flight and those of us who made it happen.

CONTENTS

ACKNOWLEDGMENTS

I would like to express my appreciation to the following people for their contributions to this book: Dr. J. D. Hunley, NASA Dryden Historian, who took on the task of reviewing my original manuscript to ensure that it was internally consistent, written for the proper audience, and historically accurate. Dr. Hunley added footnotes, expanded and edited the glossary, and compiled the present bibliography. Dr. Hunley also enlisted Dr. Darlene Lister to reorganize the original manuscript for a smooth presentation to the reading audience. To Richard P. Hallion, who is probably this country's best aviation historian, for his generous offer to allow information from his books and documents, including ***On the Frontier***, ***Test Pilots***, and ***The Hypersonic Revolution***, to be incorporated into this book. His attention to detail on facts, dates, and management of early aircraft development programs made it easier to correlate my own personal story of the lifting-body program with other activities of the time.

Also, I would like to thank Robert Kempel and Wen Painter for their generous offer to allow me to use portions of their personal story of the HL-10 Lifting Body entitled ***Developing and Flight Testing the HL-10 Lifting Body: A Precursor to the Space Shuttle***, a NASA Reference Publication published in April 1994, to be used in this book.

For their contributions to the book's technical accuracy, I would like to thank the following people who read the original manuscript and in some cases, the reorganized version, and provided technical corrections where appropriate: Bill Dana, Chuck Yeager, John Manke, and Bruce Peterson, lifting body pilots; and Robert Hoey, Bob Kempel, Wen Painter, Ed Saltzman, and Joe Wilson, lifting body engineers.

Thanks also to Betty Love, aeronautical research technician, for contributing both recorded and unrecorded flight records on original M2-F1 flights and for her comments on the reorganized version of the manuscript; John Muratore, X-38 program manager at NASA's Johnson Space Center, for reviewing the portion of the last chapter on the X-38 program; David Urie, from Lockheed, and Steve Ishmael, from NASA, for reviewing the portion of the last chapter dealing with the X-33 program; the NASA Headquarters Printing and Design office—Vanessa Nugent and Kimberly Jenkins for their Design and Editorial work, as well as Stanley Artis and Michael Crnkovic who saw the book through the printing process.

R. Dale Reed, Aerospace Engineer
NASA Dryden Flight Research Center
August 1997

Foreword

When Dale Reed asked me to write the foreword to his book, Wingless Flight: The Lifting Body Story, I had to think back a long ways to remember the day that Paul Bikle asked me to fly the M2-F1 lifting body. It was a very interesting program that would give a space vehicle similar to the present day space shuttle the ability to maneuver. During the time that the lifting body program was being flown, space capsules were re-entering the Earth's atmosphere in a ballistic path and had very little ability to maneuver.

The concept behind the lifting body program was to investigate the ability of the pilot to land in a horizontal mode which required an excessive angle of attack to flare. I enjoyed flying the lifting body and probably found it easier to fly than most pilots because of my experience with the XF-92 airplane which landed with extremely high angles of attack similar to those later experienced with lifting bodies.

Dale's book covers the warm things that go on during the test programs at Edwards Air Force Base, California. Dale has emphasized the cooperative effort that must take place between the people he calls the Real Stuff (people who create and service the flying machines) and the Right Stuff (pilots who fly the machines). Most of the NASA lifting body crews (about 90 percent) were made up of ex-military mechanics and technicians, mostly Air Force and of excellent caliber. I owe a deep debt of gratitude to many an aircraft crew chief in my career. These crew chiefs provided me with aircraft in first-class condition to fly by working themselves and their people long hours to stay on schedule.

Test pilots, on the other hand, were a different story. Dale, being a pilot himself, could see the undercurrent that flows in the macho world of test pilots. Competition has always existed between pilots. There was a special kind of competition between Air Force and NASA test pilots, and Dale has covered it very well in this book.

The lifting body story covers a little known period at Edwards Air Force Base, and it fills a gap during the transition from space capsules to maneuvering space vehicles.

Chuck Yeager
B/Gen., USAF, Ret.

INTRODUCTION

Wingless Flight tells the story of the most unusual flying machines ever flown, the lifting bodies. It is my story about my friends and colleagues who committed a significant part of their lives in the 1960s and 1970s to prove that the concept was a viable one for use in spacecraft of the future. This story, filled with drama and adventure, is about the twelve-year period from 1963 to 1975 in which eight different lifting-body configurations flew. It is appropriate for me to write the story, since I was the engineer who first presented the idea of flight-testing the concept to others at the NASA Flight Research Center. Over those twelve years, I experienced the story as it unfolded day by day at that remote NASA facility northeast of Los Angeles in the bleak Mojave Desert.

Benefits from this effort immediately influenced the design and operational concepts of the winged NASA Shuttle Orbiter. However, the full benefits would not be realized until the 1990s when new spacecraft such as the X-33 and X-38 would fully employ the lifting-body concept.

A lifting body is basically a wingless vehicle that flies due to the lift generated by the shape of its fuselage. Although both a lifting reentry vehicle and a ballistic capsule had been considered as options during the early stages of NASA's space program, NASA initially opted to go with the capsule. A number of individuals were not content to close the book on the lifting-body concept. Researchers including Alfred Eggers at the NASA Ames Research Center conducted early wind-tunnel experiments, finding that half of a rounded nose-cone shape that was flat on top and rounded on the bottom could generate a lift-to-drag ratio of about 1.5 to 1. Eggers' preliminary design sketch later resembled the basic M2 lifting-body design. At the NASA Langley Research Center, other researchers toyed with their own lifting-body shapes.

Meanwhile, some of us aircraft-oriented researchers at the NASA Flight Research Center at Edwards Air Force Base (AFB) in California were experiencing our own fascination with the lifting-body concept. A model-aircraft builder and private pilot on my own time, I found the lifting-body idea intriguing. I built a model based on Eggers' design, tested it repeatedly, made modifications in its control and balance characteristics along the way, then eventually presented the concept to others at the Center, using a film of its flights that my wife, Donna and I had made with our 8-mm home camera. I recruited the help of fellow engineer Dick Eldredge and research pilot Milt Thompson, especially in later selling the idea to others, including Paul Bikle, then the director of the NASA Flight Research Center (redesignated in 1976 the Hugh L. Dryden Flight Research Center). What followed was history, and telling for the first time in print that historic story of the lifting bodies in full and living detail is what this book is all about.

Dale Reed holding the original free-flight model of the M2-F1 filmed in 8 mm movies used to convince Dryden and Ames managers to support the program. The full-scale M2-F1 flown later is in the background. (NASA photo EC67 16475)

Between 1963 and 1975, eight lifting-body configurations were flown at the NASA Flight Research Center at Edwards AFB. They varied tremendously—from the unpowered, bulbous, lightweight plywood M2-F1 to the rocket-powered, extra-sleek, all-metal supersonic X-24B. Some configurations, such as the M2-F2, not only pushed the limits of both design engineers and test pilots but also were dangerous to fly. Film footage of the 1967 crash of the M2-F2, after test pilot Bruce Peterson lost control of this particularly "angry machine," was used about two years later as the lead-in to weekly episodes of a popular television series, ***The Six-Million-Dollar Man***, which ran for about six years. Although the M2-F2 crash was spectacular enough to inspire the concept for a popular television series, it was the only serious accident that occurred over the slightly more than twelve years of lifting-body flight-testing.

But danger has always lurked at the edge of flight innovation. All eight of these wingless wonders, the lifting bodies, were considered the flying prototypes for future spacecraft that could land like an airplane after the searing heat of reentry from outer space. The precursors of today's Shuttle and tomorrow's X-33 and X-38, the lifting bodies provided the technical and operational engineering data that has shaped the space transportation systems of today and tomorrow.

The Place and the People

The story of the lifting bodies is not just a story about wingless machines that fly. It is a story as much about people and the unique environment of the NASA facility at Edwards AFB as it is about airplanes. The driving force behind the lifting-body program was the small contingent of people at the NASA Flight Research Center at Edwards AFB in the western Mojave Desert northeast of Los Angeles.

Brought together originally in 1946 to flight-test the Bell XS-1, this little group of strong-minded individuals was also drawn to this remote facility because of their love for airplanes and the adventure of flight-testing. Being surrounded by aviation history in the making was enough to keep motivation flying high.

The NASA facility at Edwards—called initially the National Advisory Committee for Aeronautics (NACA) Muroc Flight Test Unit—was paradise to these lovers of airplanes. It was a place where people got their hands dirty working on aircraft, a place where they had the freedom to kick an airplane tire at any time. It was a place where test pilots, engineers, mechanics, and technicians all breathed the same air and walked the same halls, shops, and hangar floors. It was a place where they could take a few minutes off from tightening a bolt on an aircraft to watch a new airplane design making a flyover. The boss probably was also an airplane lover, and more than likely, he too had stopped whatever he was doing to watch the same flyover. And it was a place where about the most exciting thing in life was being involved as a volunteer in a new program.

In 1963, the lifting-body program began, circumventing the normal bureaucratic process by launching itself as a bottom-up program. It began when an enthusiastic engineer drew together a band of engineers, technicians, and pilots—all volunteers, of course—and then moved ahead, bypassing the ponderous amount of paperwork and delays of months or even years typically involved in officially initiating approved and funded aerospace programs in that era.

Besides tapping into the volunteer spirit present in the 1960s at the NASA Flight Research Center, the unofficial lifting-body program also used creative methods to locate funds. Shortly before his death in January 1991, Paul Bikle explained how that was done, saying "it was a real shoestring operation. We didn't get any money from anybody. We just built it out of money we were supposed to use to maintain the facility."[1] As the program grew over the years to involve flight-testing eight different configurations, it became more disciplined and organized. Even then, however, it was still individuals—not organizations—that made things happen.

The lifting-body concept was a radical departure from the aerodynamics of conventional winged aircraft, and it was the operational experience of the NASA and Air Force people at Edwards AFB that made the program a reality. Setting the stage for the lifting-body program was the long experience of these engineers, technicians, and pilots over previous decades in flight-testing experimental, air-launched, and rocket-boosted gliders from the XS-1 to the X-15.

A special kind of camaraderie existed among the otherwise competitive NASA and Air Force people and aircraft contractors who worked in the shops and labs of this relatively isolated facility. Often, for example, a mechanic who needed a special tool or piece of equipment would go next door on the flight line to a competing contractor and borrow what was needed. Flight-testing was difficult, demanding, and time-critical work. By helping each other get through critical times, everyone benefited from the unofficial cooperation that was a hallmark of the facility even then.

An anti-waste mentality was another hallmark of Edwards at the time. If an old piece of equipment could do the job as well as a new piece of equipment, why spend the money and time developing the new piece of equipment when the program could be moved along speedily by refurbishing and using the old one? One of the best examples of this recycling was the extensive use made of Thiokol's Reaction Motors

1. Quoted in Stephan Wilkinson, "The Legacy of the Lifting Body," ***Air & Space*** (April/May 1991), p. 54.

Division LR-11 (later designated the XLR-11), a rocket engine flown in rocket-powered experimental aircraft at Edwards for nearly 30 years, from 1947 to 1975.

The most famous use of this engine was to propel Chuck Yeager and the Bell XS-1 in the world's first supersonic flight in 1947. The Army-version LR-11 was also used to propel later models of the Bell X-1. A virtually identical Navy version called the LR-8 was used through 1959 on Douglas D-558-II rocket-powered aircraft. To keep the X-15 program on schedule, despite delays while the Thiokol XLR-99 rocket engine was being developed, a pair of old LR-11s was used in the X-15 until the bigger engine became available. During the year that followed until the XLR-99 was available, the X-15 was flown with the LR-11s and achieved speeds up to Mach 3.23. Later, many of the old LR-11 engines were donated to various aeronautical museums, some installed in the old X-1 or D-558-II aircraft and some shown as separate engine displays.

Six years afterwards, these engines were removed from the museums, refurbished, and recycled into flight-testing in the lifting-body program. Of the eight lifting-body configurations developed, four of them were powered by LR-11 rocket engines "borrowed" from museums. The last flight-test of a lifting body using an LR-11 engine occurred on 23 September 1975. Afterwards, the LR-11s found their way back to the museums, now installed in lifting bodies as well as other historic rocket-powered research aircraft.

The extremely low-cost M2-F1 launched the unofficial lifting-body program in 1963. Dubbed the "Flying Bathtub," this simple little vehicle was towed aloft by either a car or an old R4D, the Navy version of the C-47 aircraft. Except for the Hyper III, which was flown by remote control, the lifting-body vehicles were flown with research pilots on board. Two of the configurations, the M2-F2 and the first glider version of the HL-10, were marginal to control and later were modified aerodynamically to produce good flying aircraft. The original flight versions, which I call the "angry machines," tested the limits of research pilots' capabilities. We were very fortunate at the time to have a pool of the world's best research pilots to fly these marginally controllable aircraft until we, as engineers, got smart enough to convert them into good flying machines. Another lifting body, the Air Force X-24A, was converted into the X-24B, a totally new form of lifting body that I call a "racehorse" because it led toward high hypersonic aerodynamic performance.

Begun while the X-15 was still being flight-tested, the lifting-body program was unique when compared with previous research, in which most aircraft design activities were conducted by contractors and delivered to the government to meet performance specifications. For instance, the basic X-15 design, except for minor but important changes, was tested by expanding the flight envelope to the maximum speed and altitude capabilities of the aircraft. In this way, the X-15 program was mainly dri-

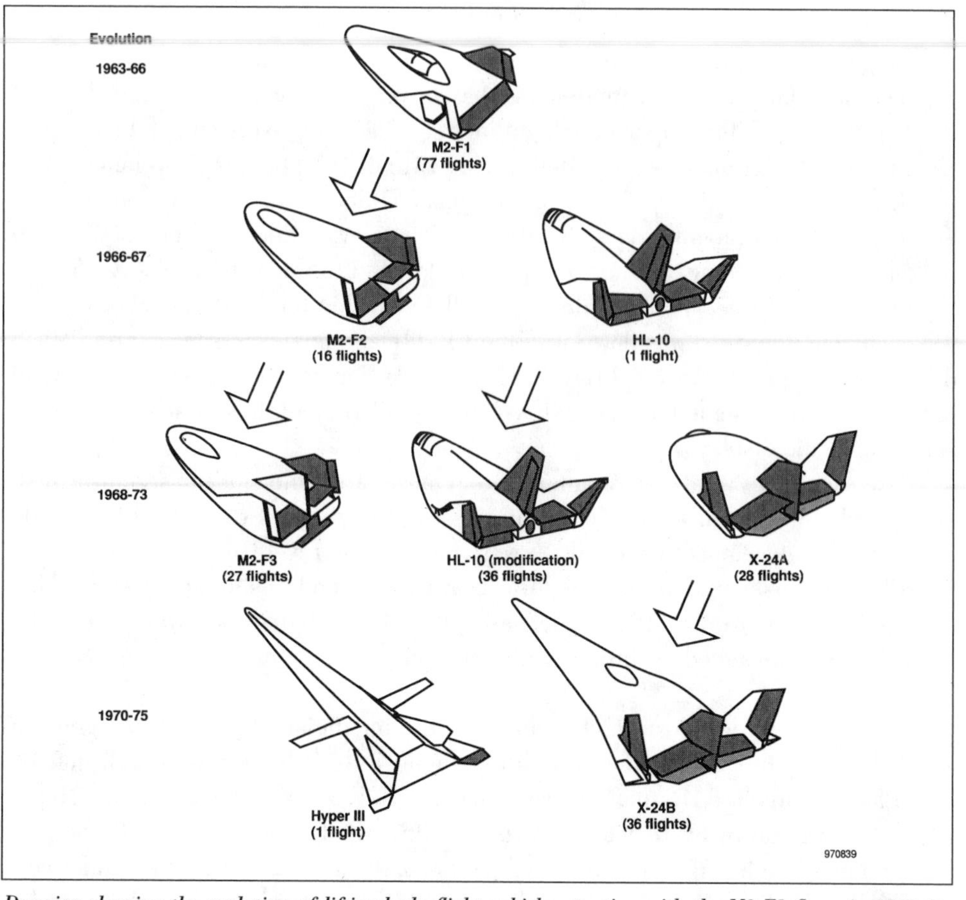

Drawing showing the evolution of lifting-body flight vehicles starting with the M2-F1 flown in 1963-66; "angry machines" M2-F2 and the original HL-10 flown in 1966-67; mature "plow-horse" lifting bodies M2-F3, HL-10 modified, and X-24A flown in 1968-73; and finally, the "race-horse" lifting bodies Hyper III and X-24B flown in 1970–75 (original drawing by Dale Reed, digital version by Dryden Graphics Office).

ven by operational and hardware considerations, whereas the lifting body was mainly a design engineer's program with NASA and Air Force engineers doing the basic aerodynamics and control-system designs, wind-tunnel testing, and simulation and control system analysis.

All of the NASA lifting-body configurations—the M2-F1, M2-F2, M2-F3, Hyper III, HL-10, and modified HL-10—were developed within NASA facilities. The aerodynamic shapes were developed in NASA wind tunnels, and the control-system control laws were developed at the Flight Research Center by NASA engineers and research pilots using simulators and other analytical techniques. Northrop, the contractor, then designed and built the hardware to meet these specifications, relying

Paul Bikle—Director of the NASA Flight Research Center from 1959 to 1971 who provided strong support for the lifting-body program. (NASA photo E68 19647)

totally on the work done by the NASA and Air Force engineers. I believe that this was an unprecedented arrangement between government and contractor technical people, everyone working together as one design team.

Paul Bikle

From what I've described so far, someone might form the impression that the NASA Flight Research Center in the 1960s was an organization of undisciplined do-as-you-like individuals. Just the opposite was true. Paul Bikle, the director of the NASA Flight Research Center at that time, was a strong disciplinarian who came to NASA from a military background. A lover of airplanes, he started his career designing light planes for Taylor Aircraft Company before World War II. He was a civilian flight-test engineer at Wright-Patterson Air Force Base, testing B-17s, B-24s, B-25s, B-29s, P-51s and other Air Force aircraft of the time. Next, he became the civilian director for flight-testing military jet aircraft with the Air Force at Edwards AFB, working closely with many top Air Force pilots, including Jimmy Doolittle and Chuck Yeager.

After his career with the Air Force, Bikle was recruited to head up the NASA Flight Research Center at Edwards, which had just been assigned to develop a flight-test program for the X-15. His ability to lead a highly disciplined flight-research organization dedicated to achieving timely results had been demonstrated many times in his Air Force career, making him an ideal choice for this job. Walt Williams—the original director of the NASA Flight Research Center—went on to lead the Mercury and Gemini space programs at Johnson Space Center.

From 1959 to 1969, Paul Bikle organized and conducted the three-aircraft, hypersonic, rocket-powered X-15 program in a highly professional and disciplined manner. Even though the X-15 program was the major activity at the NASA Flight Research Center at the time, Bikle saw the NASA facility as a research center that had to stay tuned to the aerospace world, prepared to move ahead when opportunity arose. As a result, about half of the staff was committed to X-15 research, the other half available to conduct other aeronautical research geared to the future.

Having worked closely with test pilots for years and being an accomplished pilot himself (having set the world's altitude record for sailplanes), Bikle had the uncanny ability to gauge accurately the abilities of research pilots. He also knew the abilities of most of the roughly 400 individuals then at the NASA Flight Research Center. Almost daily, Bikle wandered through the shops, talking to mechanics and engineers in their offices. Besides touring the hangars, shops, and offices, he usually played cards during lunch in the radio shop. In these ways, he stayed in touch with the pulse of the place and the people. He knew more about the daily details of the Center than did most of the engineers and project managers. He also had his own style of asking questions. He already knew the answers to the questions he was asking, but had found that asking questions was a good way of gauging how much the person knew about what was going on.

A small and balding man, Paul Bikle commanded so much respect and authority that when you met him in the hallway, he seemed ten feet tall. Years later, after he retired, he added radio-controlled model flying to his first love, soaring. One day, while he and I were flying radio-controlled gliders at the beach, I had the crazy idea that, if I had to, I could lick this friendly little guy in a fist fight. It was a crazy idea because never before had I thought of him as anything but a giant you didn't cross unless you were stupid.

Bikle disliked using up people's time with unnecessary meetings. He held one weekly meeting to take care of any and all unresolved problems. Usually, he was so attuned to daily details within the Center that he knew about a problem before it was voiced at a meeting. The meeting soon became known as the "Bikle Barrel," instilling terror in the hearts of any supervisor or project manager who had screwed up that week. Not believing that any good could come from reprimands or punishments, Bikle found that exposing screw-ups in the weekly meetings was sufficient to keep all of his people on their toes afterwards. No one was immune to the Bikle Barrel, and I had my turn a few times, too.

Bikle occasionally used other unorthodox methods to motivate people. For example, he bet several of the lifting-body people that the M2-F2 would not fly before 1 July 1966. On 8 June, the XB-70 crashed, intensifying the normal safety-of-flight worries. Even minor problems in the lifting-body program began to loom large in the aftermath of the XB-70 accident. In the next weekly meeting, Bikle decided that the entire lifting-body project would stand down for 30 days, with no attempts made to fly until all problems had been fully evaluated. At the end of the meeting, a pile of money began accumulating in front of Bikle as those with whom he had bet paid off. He simply smiled, picked up the money, and left the room. The moral: Never bet against someone who controls the game.

His more personable side came out in informal one-on-one sessions. Like most of us at the Center in those days, Bikle was in love with airplanes and loved to swap flying stories or talk about new airplane designs. Many of the big names in aviation were his personal friends. I can remember finagling my way into sitting at the same NASA cafeteria table with Paul Bikle and Chuck Yeager, just to be able to listen to them swap flying stories. In those days, I felt like a child listening to the bigger boys talk, often having to work to keep my eyes from bugging out and my mouth from dropping open in pure amazement.

Bikle was also very knowledgeable about flight-test and research techniques, even doing a professional-level flight program of his own on weekends of many of the state-of-the-art sailplanes of the time. He published their gliding performance results in reports still used today by designers of subsonic aircraft requiring very high lift-to-drag ratios.

Innovation is a personality characteristic, Bikle believed, not something that can be taught in schools or training programs. He knew that this characteristic might lie within any technician in the shop or any engineer in the office. While wandering through shops and offices, talking to various individuals, he was able to calibrate many personalities and get a feeling for individual skill levels. The door to his office was always open to anyone who had an idea that he or she wanted to share with him.

The Lifting-Body Pilots

Paul Bikle emphasized teamwork, making it clear that each engineer and technician was just as important as each research pilot to the success of the flight project. In actuality, however, the work team didn't always see it this way. The research pilots were often thought to be like the Greek gods on Mount Olympus. After all, the success or failure of a project—after long weeks, months, or years of the team's hard work—depended on one pilot doing the job right for the few minutes of that first critical flight.

Many of us involved in the project were also private or amateur pilots imbued with tremendous admiration for our fellow team members, the research pilots. Many of us envied these pilots, often trying to mentally put ourselves into their minds and bodies during flight tests. In the early days, before flights were conducted from control rooms, the radio was the primary contact point between the pilots and others on the ground. If a pilot chose to say nothing during a flight, we fairly much had to wait for the post-flight debriefing to hear how things had gone during the flight. However, we did have on-board aircraft data recordings that we could process to verify the accuracy of pilot reports.

Later, when we developed a control room at the Center for the X-15 project, research and flight-test engineers could participate in the flight by watching data displayed on consoles in various forms—dials, wiggly lines on paper rolls, and pens moving across radar maps to show the position of the aircraft. Sometimes we could influence the course of the flight by sending a message to the pilot over the radio through a control-room communicator, usually another research pilot. The ground-based communicator, who had the only radio mike in the control room, could filter comments by engineers, deciding whether they were important enough to communicate to the airborne research pilot.

As engineers, we began to feel that we were a part of the flight once we were able to see real-time data coming into the control room by way of telemetered radio signals.

Nevertheless, the spotlight remained on the research pilot. He was the man of the hour, all eyes watching to see that he did his job properly. All of the lifting-body pilots, with the exception of Chuck Yeager, had college degrees in engineering or physics. These "tigers of the air" did not fit any one stereotype, the spread of personality types ranging from the "intellectual," as represented by Fred Haise and Einar Enevoldson, to the talented "stick-and-rudder men," represented by Chuck Yeager and Joe Engle.

The flight performance of any pilot on any given day depended not only on his experience and skills but also on a number of personal factors, including whether he had had a disagreement the night before with his wife. All but one of the lifting-body pilots were current or former military fighter pilots, and fighter pilots by nature seemed to need sizable egos to be good at what they do. The spotlight appealed differently to each pilot's ego, with varying results.

For example, some of the lifting-body configurations had very poor flying characteristics, which created situations in which pilots could cause oscillations by overcontrolling. This condition is called "pilot-induced oscillation" (PIO), a deviation from controlled flight that can happen with the best of pilots if the flying characteristics of the aircraft are bad enough. However, the pilots with the biggest egos often had the most difficulty admitting they were involved in a PIO situation during a flight.

The lifting-body pilots also seemed to belong to an unofficial but exclusive club in the pilots' office. The performance of any pilot could be judged only by his fellow pilots or by his boss, Paul Bikle for the NASA pilots and various Air Force commanders for the Air Force pilots. It was not considered proper for flight-test or research engineers to suggest that a pilot's performance was not up to par. The lifting-body pilots included many top test pilots. Consequently, problems in flying the lifting-body vehicles were often thought to be the fault of the engineers who had created configurations that were marginally controllable, rarely if ever considered to result from any lack of piloting skill.

Chuck Yeager had his own pilot rating system, the pilot bosses had theirs, and we research engineers had our own. As research engineers, we unofficially divided the pilots into two categories: those who were research test pilots, who would try hard to bring home quality data, and those who were just test pilots, who could expand envelopes and bring the aircraft home safely but who were sloppy with regards to data. We were fortunate that most of the lifting-body pilots were also true research test pilots and that we got the data we wanted.

The era of the lifting bodies began with a very modest program involving only one pilot, Milt Thompson. The program grew over the years to include eight different lifting-body configurations flown by 17 pilots, eight of whom were NASA, the others Air Force. Sixteen of the seventeen pilots had fighter aircraft backgrounds and one, Dick Scobee, had large airplane experience.

	Number of flights								
Pilot	M2-F1	M2-F2	HL-10	HL-10 modified	M2-F3	X-24A	Hyper III	X-24B	Total
1. Milt Thompson	45	5					1		51
2. Bruce Peterson	17	3	1						21
3. Chuck Yeager	5								5
4. Don Mallick	2								2
5. James Wood	Car T.								
6. Don Sorlie	5	3							8
7. Bill Dana	1			9	19			2	31
8. Jerry Gentry	2	5		9	1	13			30
9. Fred Haise	Car T.								
10. Joe Engle	Car T.								
11. John Manke				10	4	12		16	42
12. Pete Hoag				8					8
13. Cecil Powell					3	3			6
14. Mike Love								12	12
15. Einar Enevoldson								2	2
16. Francis Scobee								2	2
17. Tom McMurtry								2	2
Total	77	16	1	36	27	28	1	36	222

970830

Lifting-body pilot list showing numbers of flights per lifting body by each of the 17 lifting-body pilots (compiled by Betty Love).

All of the pilots had other test or research responsibilities on other aircraft programs within NASA and the Air Force, the typical lifting-body flights being weeks or even months apart. Often, these other programs involved research or developmental military aircraft being tested at Edwards at the same time we were flying the lifting bodies. We were fortunate in the lifting-body program to be able to tap into this elite source of pilots when we needed them.

We were even able to get Chuck Yeager to take time from his busy schedule during the first year of the lifting-body program to fly the M2-F1 and give his assessment of this vehicle. Three of the lifting-body pilots went on to be astronauts. Fred Haise went to the Apollo program and flew the Shuttle landing approach tests. Joe Engle and Dick Scobee became Shuttle commanders for space flights.

A total of 222 lifting-body flights were made in those twelve busy years. Topping the list was the M2-F1 with 77 air tow flights. The HL-10 Modified and the X-24B had 36 flights each. The X-24A flew 28 times; the M2-F2 had 16 flights; the M2-F3, 27; and the original HL-10 and Hyper III had only one flight each.

Here is a thumbnail introduction to the pilots, given in the order in which they first flew vehicles in the lifting-body program:

Milton O. Thompson, the first lifting-body pilot, flew the M2-F1 on its first flight on 16 August 1963. Milt flew the M2-F1 16 more times before the next two pilots,

Milt Thompson—first lifting-body pilot—standing beside the M2-F1 configuration selected for flight (without a center fin). (NASA photo EC63 206)

A happy Bruce Peterson—second lifting-body pilot—after he successfully piloted the marginally controllable HL-10 on its first flight. (NASA photo E66 16199-1)

Bruce Peterson and Chuck Yeager, were invited to fly it. In all, Milt flew the M2-F1 45 times. He also made the first five flights of the heavy-weight M2-F2 lifting body, a grand total of 51 lifting-body flights. All of his flights were glide flights.

Milt was instrumental in the start-up of the lifting-body program. It would have been difficult to sell the lifting-body program to project managers without the help of Milt's charm. After flying the M2-F2, Milt retired as a flight research pilot, then moved into setting up training programs and working with Paul Bikle in evaluating new pilots for the later lifting-body projects.

Bruce A. Peterson, the second lifting-body pilot, made a total of 21 flights on three different lifting bodies: the M2-F1 17 times, the M2-F2 3 times, and the HL-10 once. On 22 December 1966, he became the first pilot to fly the HL-10. He retired from test flying following the crash of the M2-F2 on 10 May 1967.

Chuck Yeager was the third pilot to fly a lifting body, making five flights of the M2-F1, one on 3 December 1963, and two each on 29 and 30 January 1964. Paul Bikle wanted his old friend and master test-pilot, Colonel (later General) Chuck Yeager, to fly the M2-F1 early enough to give an assessment before other Air Force pilots flew the vehicle. At the time, Yeager headed up the USAF Aerospace Research Pilot School, also known as the Test Pilot School, at Edwards. Bikle thought that Yeager gave the most accurate and descriptive flight test report of any pilot that Bikle had ever worked with in the Air Force or NASA. Although Yeager never flew any of the rocket-powered lifting bodies, he exerted considerable influence, encouraging the Air Force in developing the rocket-powered X-24A and X-24B as well as in the conceptualization of the jet-powered X-24J, which was never built.

Yeager could be very blunt and straightforward when it came to evaluating the performances of other test pilots, and perhaps those who received the brunt of his criticism might not hold him in as high a regard as I and others do. Yeager basically divided test pilots into two categories: those who can hack it, and those who cannot. He minced no words in his verbal or written criticism of those pilots who made more than a limited number of mistakes in the stick-and-rudder department. Nor did he mince words in evaluating how well an aircraft handled or performed.

The fourth lifting-body pilot, Donald L. Mallick made only two lifting-body flights with the lightweight M2-F1 on 30 January 1964. James W. Wood, the fifth lifting-body pilot, made only car tows on 6 February 1964. Major Wood was transferred by the Air Force to another command and did not get a chance to fly the M2-F1 in air tow. He had been one of the original X-20 (Dyna-Soar) pilots selected by the Air Force.

Donald M. Sorlie, the sixth lifting-body pilot, made his first air-towed flight in a lifting body on 27 May 1965. The official Air Force "boss" of the lifting-body and X-15 Air Force test pilots, Lieutenant Colonel Sorlie made five flights in the M2-F1 and three in the M2-F2, just enough to evaluate what kind of challenge would con-

Bill Dana, seventh lifting-body pilot, who flew 4 lifting-body configurations including 19 flights on the M2-F3 for a total of 31 lifting-body flights. The HL-10 is shown behind him. (NASA photo E69 20288)

Then-Capt. Jerauld Gentry, principal Air Force and eighth lifting-body pilot overall, who flew 5 lifting-body configurations including 13 on the X-24A for a total of 30 lifting-body flights. (NASA photo, EC97 44183-1)

front his test pilots in these lifting bodies. At the time, he was Chief of the Fighter Operations Branch, Flight Test Operations—-the primary pool of Air Force test pilots at Edwards AFB.

The seventh pilot to fly a lifting body, William H. Dana, had 31 lifting-body flights over a little more than ten years, flying the lifting-bodies over a longer span of time than did any other pilot. He had his first lifting-body flight in the M2-F1 on 16 July 1965. He also flew the HL-10 and the M2-F3. His last lifting-body flight was in the X-24B on 23 September 1975.

Dana received the NASA Exceptional Service Medal for his ten years as a research pilot in four of the lifting-body vehicles (M2-F1, HL-10 modified, M2-F3, and X-24B). In honor of his research work on the M2-F3 lifting-body control systems, Dana in 1976 received the American Institute of Aeronautics and Astronautics' Haley Space Flight Award.

Jerauld R. Gentry, the eighth lifting-body pilot, was the chief Air Force lifting-body pilot, making a total of 30 lifting-body flights. Major Gentry made his first air-towed flight on the M2-F1 on 16 July 1965. He also flew the M2-F2, the HL-10, and the X-24A. He made his last lifting-body flight in the M2-F3 on 9 February 1971. Major Gentry developed a reputation as an outstanding lifting-body research pilot, flying the X-24A on its first glide flight as well as its first rocket-powered flight, demonstrating a high level of skill in gathering the flight data needed by engineers in expanding the X-24A's flight envelope.

The ninth and tenth lifting-body pilots, Fred Haise and Joe H. Engle, flew the M2-F1 on car tows up to altitudes of 25 and 30 feet on 22 April 1966. Neither of them flew the M2-F1 from airplane tows, nor did they fly any of the B-52 launched lifting bodies.

Soon after flying the M2-F1 in 1966, Haise was assigned as an astronaut at what became the Johnson Space Center, precluding any additional involvement with the lifting-body project. Later, Haise was on the ill-fated Apollo 13 flight, which almost ended in disaster following an explosion in space, the topic of the popular movie ***Apollo 13*** that premiered in 1995. General Joe Engle also had his assignment to the lifting-body project cut short when he was one of 19 astronauts selected in March 1966 for NASA space missions. I would have liked to have seen how well Joe Engle, in particular, would have performed over time as a lifting-body pilot. He shared many of Chuck Yeager's characteristics: he, too, was full of 'piss and vinegar' as well as one of the best stick-and-rudder men around.

John A. Manke, the eleventh lifting-body pilot, was the second busiest with 42 lifting-body flights, the busiest being Milt Thompson with 51. Most of Manke's flights were rocket-powered, while all of Thompson's were glide flights, including the remotely piloted Hyper III in which Milt "flew" from a ground cockpit. Manke's first flight

John Manke, eleventh lifting-body pilot, who flew 4 different configurations including 16 X-24B flights for a total of 42 lifting-body flights. (NASA photo EC69 2247)

Lt. Col. Michael Love, fourteenth lifting-body pilot, who flew the X-24B 12 times. (NASA photo E75 29374)

was a glide flight on the modified HL-10 on 28 May 1968. Manke flew the HL-10 ten times, the M2-F3 four times, the X-24A twelve times, and the X-24B sixteen times. He made his last flight on 5 August 1975 in the X-24B.

The twelfth pilot to fly a lifting body, Peter C. Hoag, first flew the modified HL-10 on 6 June 1969. Major Hoag made his eighth flight on the HL-10 on 17 July 1970. This was also the last flight of the HL-10. While flying the HL-10 on 18 February 1970, Major Hoag set the speed record for all of the lifting bodies—Mach 1.86.

Cecil William Powell, the thirteenth lifting-body pilot, had his first lifting-body flight on 4 February 1971, a glide flight in the X-24A. He flew the X-24A and the M2-F3 three times each. His last flight on a lifting-body was a rocket flight on the M2-F3 on 6 December 1972.

Fourteenth among the pilots to fly a lifting body, Michael V. Love first flew the X-24B on 4 October 1973. A year later, on 25 October 1974, Lieutenant Colonel Love set the speed record of Mach 1.75 for the X-24B. On 20 August 1975, he had his twelfth and final flight of the X-24B.

Einar Enevoldson, the fifteenth lifting-body pilot, made his first of two glide flights in the X-24B on 9 October 1975. He was one of three guest pilots invited to fly the X-24B in glide flights as part of a guest-pilot evaluation test exercise at the end of the

X-24B flight program after the official research flights had been completed. Each of the three guest pilots (including Major Francis R. "Dick" Scobee and Thomas C. McMurtry) flew the X-24B twice.

Francis R. "Dick" Scobee, the sixteenth lifting-body pilot, first flew the X-24B on 21 October 1975. Primarily an Air Force transport test pilot, Major Scobee was the only lifting-body pilot with no background as a fighter pilot. He kidded us, saying he was selected as a guest pilot to prove that if a transport pilot could fly the X-24B, then any pilot could fly future spacecraft versions of the X-24B.

The X-24B shared very similar speed and performance characteristics with the projected Shuttle spacecraft design, so the X-24B was used to collect operational data used in the design and development of the Space Shuttle vehicles. Scobee said that his experience flying the X-24B inspired him to apply to the NASA Astronaut Corps to fly the Shuttle spacecraft. He was selected as an astronaut for NASA in January 1978. On 28 January 1986, Scobee unfortunately perished in the ***Challenger*** explosion.

Thomas C. McMurtry was the seventeenth and final pilot to fly a lifting body, doing so as the third invited guest pilot at the end of the X-24B program. He flew the X-24B in glide flight twice, once each on 3 and 26 November 1975.[2]

How *Wingless Flight* Came to be Written

My life-long love affair with airplanes has kept me from truly retiring. After I retired from NASA in 1985, I was recruited to manage the development at Lockheed of various Remotely Piloted Vehicles (RPV), working four years at the Lockheed Advanced Development Plant known as the "Skunk Works," managing design, vehicle development, and flight-test programs.

After I left Lockheed in 1989, still unable to pull myself away from an active involvement with aircraft, I served as a consultant to various aircraft organizations and soon found myself working as a contractor, supporting NASA programs at NASA Dryden, Edwards AFB. Once more I was able to work with some of my old NASA friends at Dryden, including Milt Thompson.

Milt had been working on a book entitled ***At the Edge of Space***,[3] which told the story of the X-15. After this book was published in 1992 by the Smithsonian Institution Press, Milt was asked if he would write a book telling the lifting-body story. For several years, I had thought of writing just such a book. However, at the time, I

---

2. Thanks to Betty Love for checking and correcting the statistics for this section.

3. Milton O. Thomposon, ***At the Edge of Space: The X-15 Flight Program*** (Washington, DC: Smithsonian Institution Press, 1992).

was too busy having fun coming up with new ideas for creating new airplane programs. With the new miniature computers and global-positioning satellite systems, I was totally involved with developing autonomously-controlled unpiloted air vehicles of all sorts.

Milt Thompson died suddenly on 6 August 1993. Before his death, Milt had begun writing the book that would tell the lifting-body story, but he had not finished it at the time of his death, leaving me as the only remaining lifting-body team member who knew the full lifting-body story from beginning to end. If a book telling the entire story were to be written, it seemed that I was the only participant left who could do it.

By this time, the professional aerospace writer and historian Richard P. Hallion had already published three excellent histories telling aspects of the story. First published by the Smithsonian in 1981, ***Test Pilots***[4] tells the complete story of flight-testing, from the earliest tower jumps in 1008 to the around-the-world flight of the Voyager in 1986. ***On the Frontier***,[5] published in 1984 as a volume in the NASA History Series, is a comprehensive history of flight research at NASA Dryden after World War II, 1946-1981. ***The Hypersonic Revolution***,[6] published in 1987 by the U.S. Air Force, is mammoth in scope, covering events from 1924 to 1986—from the early rocket experi-ments to the aerospace plane.

Richard Hallion has already done an excellent job in these books in documenting the historic facts as well as the political and managerial aspects of the lifting-body story. What remains untold is the story that facts alone cannot tell: the human drama as it unfolded in the day-by-day activities of the people who lived and breathed the lifting-body adventure from 1963 to 1975.

Wingless Flight tells that story, for I remain convinced that it is about more than machines; it is at least as much about the people with the "real stuff," who created and maintained the machines, as it is about the individuals with the "right stuff," the pilots who flew the lifting bodies.

---

4. Richard P. Hallion, ***Test Pilots: The Frontiersmen of Flight*** (Washington, DC: Smithsonian Institution Press, 1992).

5. Richard P. Hallion, ***On the Frontier: Flight Research at Dryden, 1946-1981*** (Washington, DC: NASA SP-4303, 1984).

6. Richard P. Hallion, ***The Hypersonic Revolution: Eight Case Studies in the History of Hypersonic Technology*** (2 vols.; Wright-Patterson Air Force Base, Ohio: Special Staff Office, 1987). Since these lines were written, another study of Dryden history appeared, Lane E. Wallace's ***Fights of Discovery: 50 Years at the Dryden Fight Research Center*** (Washington, DC: NASA SP-4309, 1996). Based on an earlier version of Wingless Flight and an interview with Dale Reed, this short history devotes considerable attention to the lifting-body story.

CHAPTER 1

THE ADVENTURE BEGINS

My journey in February 1953 to the NACA High Speed Flight Research Station (as the Muroc Flight Test Unit had come to be called in 1949) actually began about a decade earlier in two small mountain towns in Idaho, about as far from the center of aerospace innovation as one can get. My roots are with farmers and ranchers, my grandfather having moved his family members from Kansas to the sagebrush country of southern Idaho to carve out their future in agriculture, both of my parents the children of farming families.

Around age twelve, I was smitten with what would prove to be a lifelong love of airplanes. I still remember the summer day when I saw my first sailplane. John Robinson had come to Ketchum, Idaho, with his one-of-a-kind sailplane called the Zanonia to try for some world sailplane records. A beautiful craft, the Zanonia had gull wings reminiscent of some of the German sailplanes of the time. Robinson cleared the brush from a flat area across the road from my family's home, making a small dirt strip. Here, Robinson would use a car to tow the Zanonia aloft, the sailplane rolling on a dolly with a set of dual wheels that would drop by parachute after take-off.

For two weeks that summer, I helped Robinson, untangling the tow-line from the brush after the glider had been launched and picking up the parachuted landing gear. I loved to lie on the grass, watching the Zanonia riding the air currents around the mountain peaks. Robinson set two world altitude records in the Zanonia that summer, flying the waves and thermals above the Sawtooth Mountains.

I then began building and flying model gliders and free-flight model airplanes. A hundred miles stood between me and the next modeler in those days, so I was fairly much on my own, except for some occasional help from my mother who was good with crafts and taught wood shop at the local grade school. Fairly quickly I learned I had to limit the duration of my engine runs, else chance losing my models when they glided down on the other side of the hills or mountains.

One September day, one of my models did exactly that. It caught a thermal and flew over a nearby mountain. Two weeks later, my father found that model perched unharmed on a bush at the bottom of a gully two miles from the ridge it had flown over. I flew that model for another year, during which I equipped it with floats so it could fly off of a nearby mountain lake.

Across the street from my high school in Hailey, about 12 miles south of Ketchum, was a grass field where a bush pilot-operator named Bob Silveria kept two airplanes. During the summer and fall months, the big radial engine of his old Waco cabin biplane could be heard lumbering through the Sawtooth Mountains, carrying fishermen and hunters to the primitive wilderness landing strips along the Middle Fork of the Salmon River. Silveria also had a 65-horsepower Aeronca Defender L-3 airplane at the grass field, using it to give flying lessons as well as to transport hunters into the flats south of Hailey where they chased coyotes.

By the time I was sixteen, even my high school physics and chemistry teacher, Mr. Kinney, knew that I was interested in airplanes. A private pilot who was good friends with Silveria and occasionally rented the little Aeronca airplane across the street from the school, Mr. Kinney offered to teach a class in aeronautics if I could round up eight interested students. I found six interested boys fairly easily, but I had to overcome my shyness around the opposite sex long enough to talk two girls into joining us to fill the class.

Mr. Kinney used the little Aeronca as a teaching tool. We learned to hand-prop to start the airplane and taxi it around the grass field. We did everything but fly. Seeing my enthusiasm, Mr. Kinney encouraged me to apply for a student license and take some flying lessons. I did not know that his suggestion was part of a plot hatched between him and Silveria to see how soon they could get me to solo.

On a cool September day in my sixteenth year, I had my first flying lesson. As I sat in the front seat of the Aeronca, Silveria told me that my job was to handle the throttle, rudder pedals, and brakes, that he would do everything else with the stick from the back seat. All I had to do was put my hand lightly on the stick and follow his movements.

Since Mr. Kinney had earlier done a good job in teaching me in the class on how to taxi a tail-wheel airplane, I had no problem when Silveria told me to set the trim, taxi to position, and start the takeoff run. I knew that my task was simply to steer the rudder pedals and touch but not move the control stick. As we rolled across the grass field, the tail came up eventually and we rolled along on two wheels. I remember thinking what a smooth pilot Silveria was, for I hadn't noticed any movement at all on the stick. Soon we were flying, but I still hadn't noticed any movement on the stick. We had climbed to an altitude of 500 feet when Silveria, his first words to me since the takeoff, said, "Do you know that you made that takeoff by yourself without my help?"

I couldn't believe it, for I was doing practically nothing to fly the airplane. All I had done was make very small and gentle inputs to the rudder while we were on the ground and once we were in the air. I think I made those small control inputs automatically, perhaps subconsciously, because I had learned from building and flying model airplanes that a properly designed airplane can do a pretty good job of flying, even without the pilot.

A few days later, after three and a half hours of flight instruction, I soloed. By age sixteen, then, I was totally hooked on aviation. At first, I thought I wanted to be a bush pilot in Alaska or somewhere else equally exciting, but my high school principal talked me into going to college and studying engineering. Off I went to the University of Idaho in Moscow. Unlike other universities at the time, Idaho didn't offer a major in aeronautical engineering, but all I could afford was Idaho. I majored in mechanical engineering, taking as many aeronautical courses as I could.

Little better than an average student in high school, I found myself getting almost straight As in college. I had found my niche in aeronautical engineering, thanks to a love of flight and airplanes that had begun when I was only twelve, a young boy in a small mountain town in Idaho, far away from the center of aviation's innovative future.

As I took college courses, I found myself more and more intrigued by what I was reading in magazines about what was happening at Edwards AFB in Southern California, where a small contingent of NACA people was flight-testing the world's first supersonic airplane, the rocket-powered X-1. Little did I know, as I read these articles, that soon I would be a part of that small contingent of NACA people, conducting my own aeronautical experiments on the X-1 and becoming personally acquainted with the famous test pilot Chuck Yeager.

Before leaving Idaho in early 1953 to report to work at the High Speed Flight Research Station in the Mojave Desert, I did some reading on the history of the NACA and the Mojave Desert site. And then I got into my car, drove south from Idaho and west across the Nevada desert to the town of Mojave, California, where I made a sharp southeastern turn into the middle of nowhere.

At that time, Edwards Air Force Base was very small and compact, located on the edge of Muroc Dry Lake, now known as Rogers Dry Lake. The name of the base had changed only a few years earlier from Muroc Army Airfield to Edwards Air Force Base in honor of Captain Glen W. Edwards, killed in June 1948 in the crash of a Northrop YB-49, an experimental flying wing bomber.

In late 1946, the NACA had sent thirteen engineers and technicians from the NACA Langley Memorial Aeronautical Laboratory to Muroc Army Airfield to assist in flight-testing the Army's XS-1 rocket-powered airplane. These thirteen individuals fairly much made up what was soon to be called the NACA Muroc Flight Test Unit. Over the next fifty years, the NACA Muroc Flight Test Unit grew into what is today the NASA Dryden Flight Research Center with over 900 NASA employees and contractors supporting NASA's premiere flight-test activities.

Ground Zero: The Place Where Tomorrow Begins

The flight-testing of all experimental and first-model military aircraft occurred along an ancient dry lake now called Rogers Dry Lake, located on the western edge of California's Mojave Desert just south of Highway 58 between the towns of Boron and Mojave. Only a few miles northeast is the world's largest open-pit borax mine. Within sight of Rogers Dry Lake is one of the first immigrant trails through California.

The original name of the site, the NACA Muroc Flight Test Unit, comes partly from local history. "Muroc" is "Corum" spelled backwards. The first permanent settlers in the area, the Corum family located near the large dry lake in 1910. Later, they tried to get the local post office named Corum. However, there was already one with a nearly identical name (Coram) elsewhere in California, so they reversed the letters to spell Muroc.[1]

What was there about this dry lake that made it ideal as the later site of major aviation flight-test history? About 2,300 feet above sea level, Rogers Dry Lake fills an area of about 44 square miles—nearly half again as large as New York's Manhattan Island—and its entire surface is flat and hard, making it one of the best natural landing sites on the planet. The arid desert weather also provides excellent flying conditions on almost every day of the year.

Rogers Dry Lake is the sediment-filled remnant of an ancient lake formed eons ago. Several inches of water can accumulate on the lakebed when it rains, and the water in combination with the desert winds creates a natural smoothing and leveling action across the surface. When the water evaporates in the desert sun, a smooth and level surface appears across the lakebed, one far superior to that made by humans.

Water on the surface of Rogers Dry Lake also brings to life an abundance of small shrimp—several unique species of the prehistoric crustacean—but they disappear once the desert sun evaporates the water. Annual rainfall here is only about four to five inches, considerably less in some years. In extremely wet years, the annual rainfall can rise to six or even nine inches.

Winds are quite predictable, usually from the southwest during spring and summer, with a mean velocity of six to nine knots. Sunrises and sunsets can be breathtakingly beautiful, as can the spring wildflowers with enough rain.

The surrounding area is typical of the California high desert with rolling sand hills and rocky rises, ridges, and outcroppings punctuated in the low spots with dry lakebeds. Mountains lie on three sides—at the south, west, and north—with the mighty Sierra Nevada range to the north rising to over 14,000 feet. Joshua trees cluster among the chaparral and sagebrush. A type of Yucca (a member of the Lily family), the Joshua tree has clusters of very sharp and dark green bayonet-shaped or quill-like spines that grow six to ten inches long and that only a botanist would call "leaves." Like everything else in the surrounding desert, the Joshua tree is well suited for survival in a harsh environment. In summer, temperatures can reach or exceed 120 degrees Fahrenheit, with 10 to 15 percent humidity. In winter, temperatures can fall to nearly 0 degrees Fahrenheit.

In a curious coincidence, two entirely different and likely unrelated men named Joe Walker figure prominently in local pioneering history, separated by about 115 years. In the spring of 1843, Joseph B. Chiles organized and led one of the first wagon trains out from Independence, Missouri, to California. At Fort Laramie in Wyoming, he met an old friend, Joe Walker, who joined the California-bound wagon train as a guide. Once in California, the wagon train ran low on provisions and split into two groups, one on horseback led by Chiles that went north to circumvent the Sierra

1. Hallion, ***On the Frontier***, pp. xiv-xv.

Nevadas, the other in the wagons led by Joe Walker heading south. The people in the Walker party had to abandon their wagons just north of Owens Lake, arriving on foot at what is now called Walker Pass at eleven in the morning on 3 December 1843. Walker Pass, named after the first Joe Walker, is only 56 miles across the southern Sierra Nevadas from Edwards AFB where, 115 years later, another man named Joe Walker, the prominent NACA/NASA X-15 research pilot, was engaged in a very different kind of pioneering.

In the 1930s, early aviators—including the military and private airplane designers such as John Northrop—used Rogers Dry Lake as a place to rendezvous and test new designs. During World War II, the U.S. Army Air Corps conducted extensive training and flight-testing at the site. This is also the general area where a colorful social club and riding stable was located, established by the aviatrix Florence "Pancho" Barnes and frequented by many of the early and famous test pilots and notables of aviation history.

In more recent times, the Air Force, NASA, and various contractors have used Rogers Dry Lake in conducting flight tests on many exotic and unusual aerospace vehicles. In the words of Dr. Hugh L. Dryden—the early NACA/NASA leader, scientist, and engineer—the purpose of full-scale flight research "is to separate the real from the imagined . . .to make known the overlooked and the unexpected," words that help clarify why a remote location in the western Mojave Desert would become the site where innovative NASA engineers and technicians would gather to help create the future of aviation.[2]

The official name of the site has changed over the years. It changed its name from the NACA High Speed Flight Research Station to the NACA High Speed Flight Station in 1954 and then to the NASA Flight Research Center in 1959. It became the NASA Hugh L. Dryden Flight Research Center in the spring of 1976, a name it regained in 1994 after a hiatus from 1981 to that year as the Ames-Dryden Flight Research Facility. However, when I arrived at the site in 1953, it was still called the NACA High Speed Flight Research Station, and the people at the facility were conducting all of the NACA's high-speed flight research. They were used to conducting high-performance flight research on rocket-powered vehicles that had to land unpowered. Unpowered landings with high-performance aircraft became relatively routine, but not necessarily risk-free, on the vast expanse of Rogers Dry Lake.

---

2. Hugh L. Dryden, "Introductory Remarks," National Advisory Committee for Aeronautics, ***Research-Airplane-Committee Report on Conference on the Progress of the X-15 Project,*** (Papers Presented at Langley Aeronautical Laboratory, Oct. 25–26, 1956), p. xix. I am indebted to Ed Saltzman for locating this quotation, the words for which are common knowledge at the Center named in honor of Hugh Dryden but the source for which is not well known.

Flight Research, 1953–1962

When I arrived at the Station in February 1953, the purely rocket-powered Bell X-1 and X-2 as well as the Douglas D-558-II experimental aircraft were being flight-tested, air-launched from B-29s and B-50s (essentially the same as the B-29, but with slightly different engines). Before I arrived on the scene, the Air Force had operated the B-29s, but the NACA had taken over operating the B-29s by the time I got there, including the B-29 used for the D-558-II, which had the distinction of being the only Navy-owned B-29 (Navy designation, P2B). Also being flown then was a second D-558-II with a hybrid turbojet-rocket propulsion system. Other experimental turbojet aircraft being flown included the Bell "flying wing" X-4 (technically, a swept wing combined with an absence of horizontal tail surfaces), the Bell variable-sweep-wing X-5, and the first high-performance delta-wing aircraft, the Convair XF-92.

At the NACA facility at that time, all new junior engineers were expected to learn the flight-research business from the bottom up. Given the limited data systems of that era, plus the lack of high-speed computing capability, a research engineer's job was about ninety percent measuring and processing data and only about ten percent analyzing and reporting the flight results. With all the weird and wonderful airplanes at that time, stability and control problems were prominent. Most of the senior engineers at the NACA facility were busy analyzing and trying to solve these problems. This meant there was a lot of pick-and-shovel work for the junior engineers to do.

My first job assignment involved measuring aerodynamic loads on the wings and tail surfaces of various research aircraft. Hundreds of strain gauges had been installed inside the structures of these aircraft as they were built in the factory. My task was to calibrate these gauges and other data acquisition devices on the aircraft, including control position indicators, air data sensors, gyros, and accelerometers.

Today, these tasks are the responsibility of instrumentation engineers. Earlier, due to the small staffing at the NACA facility, these tasks fell on the shoulders of the aero or research engineers. One advantage back then of doing things this way was that by the time research engineers finally had enough flight data to analyze, they had good knowledge of the accuracies of the instrumentation—so good that if weird glitches turned up in the data during flight tests, they were better able to determine whether the data was real or indicated a problem in the instrumentation.

My first task involved measuring the aerodynamic loads on the X-5 research aircraft, a little airplane that had evolved from a design smuggled out of Germany at the end of World War II. Bell Aircraft completed the design, building what was to become the world's first variable-wing-sweep aircraft.

My task involved measuring the bending, shear, and torque loads of the wing and tail surfaces on three configurations of the X-5, one each with 20-, 40-, and 60-degree wing-sweep angles. This meant that I had to have separate wing strain gauge calibrations for each wing sweep. In those days, calibrations involved manual labor at about thirty load points on each wing and tail surface. I spent long hours over days and even weeks on a jack handle, putting incremental loads on the airplane at all of these points.

Each strain gauge output was read off a meter and written down by hand, resulting in stacks of paper with handwritten data, then processed by hand on the old mechanical Frieden calculating machines. Processing involved selecting groups of multiple strain gauges and developing equations for bending moments, shear and torque.

A staff of ten women did all the calculations on the Frieden machines. To me, they seemed the hardest working people at the facility, each of them spending long hours clanking away on a calculating machine. In those days, we worked in old barracks-type buildings with swamp coolers on the windows. The Frieden machines had to be carefully covered up during desert dust storms, for the dust coming through the coolers could ruin those mechanical wonders.

Over the years, I became a specialist in this sort of measurement work, doing flight research with the X-1E, F-100A, D-558-II, and X-15. During the X-15 program, my area of expertise expanded into aerodynamic heating, and my responsibilities grew to include each planned X-15 flight as speeds and altitudes increased far beyond those for any existing aircraft.

As the X-15 pushed closer and closer to its maximum speed of Mach 6.7 (or 6.7 times the speed of sound) and maximum altitude of 354,200 feet (or 70 miles above the earth), the Inconel-X steel and titanium structure of the X-15 could reach temperatures as high as 3,000 degrees Fahrenheit in areas of concentrated aerodynamic heating. The X-15 had been instrumented with hundreds of strain gauges and thermocouples for measuring the stresses and heat in its structure.

North American Aviation's structural designer of the X-15, Al Dowdy, and I worked as a team examining each planned flight to determine if there was any cause for concern about structural failure. For each flight, the flight-planning team and the pilot would develop a flight plan on the facility's X-15 flight simulator. With this planned flight profile of speed, altitude, angle of attack, and load factor, we could calculate aerodynamic heating inputs to the external skin on various parts of the aircraft.

Al and I selected seven critical areas of structure on the X-15 to monitor in detail during each flight program. For example, one wing area included the Inconel-X steel skin and the titanium spar caps and webs. From the information gained from monitoring these areas, I could then generate time histories of the temperature rise and decline in each element of the aircraft's structure. With this data, Al, in turn, could determine the stresses within the structure by combining calculated aero loads with my calculated temperatures. At flight-planning tech briefings, I would then report whether I thought the planned flight would be safe from the structural standpoint.

Throughout the X-15 program, we continued to test our prediction techniques by comparing our preflight calculations with measured temperature data from the actual flight. For some skin areas, we revamped our calculations to include laminar heating when we thought that it would be turbulent heating, and vice versa. Turbulent heating results in temperatures almost twice as high as those due to laminar heating, but at first we weren't always smart enough to know whether the flow would be laminar or turbulent. Later, we became more skilled at predicting external aerodynamic heating. We still had to refine our calculations for internal heat flow because the structural joints did not transfer the heat as anticipated. We determined the correction factors for the heat-transfer equations while the X-15 was still flying at low supersonic speeds of up to Mach 3. By the time that structural heating became more critical near Mach 5 and Mach 6 later in the X-15 program, we could do a much more accurate job in predicting structural temperatures.

The Early 1960s: Concepts of the Lifting Body

Although I gained a great deal of satisfaction as a researcher in structures on the world's first hypersonic airplane, my interests in aeronautics always had been much broader than aircraft structures. Still very much interested in stability and control, aerodynamics, and unusual aircraft configurations, I continued to design and build model airplanes and to fly light planes and sailplanes on my own time.

I was also fascinated with the space program, following closely the activities of Walt Williams, my first boss at the facility, and the people he took with him from the NASA Flight Research Center to Johnson Space Center to conduct the Mercury and Gemini programs. While reading NASA and Air Force reports on design concepts for future spacecraft, I noticed a pattern developing. Although many of these studies included concepts of lifting reentry vehicles, when actual space vehicles were designed, they were always non-lifting or ballistic capsule-type vehicles.

At the time, it was obvious that NASA and Air Force decision-makers had little confidence in the concept of lifting reentry and even less for lifting-body types of reentry vehicles. Although it funded many studies of lifting reentry configurations of all types, including lifting bodies, the Air Force soon concluded that lifting bodies were too risky.

In September 1961, a blue-chip panel of the Scientific Advisory Board chaired by Professor C.D. Perkins had recommended to Air Force General Bernard A. Schriever that all expenditures on flight hardware be made solely for winged vehicles, not lifting bodies. The panel had questioned the control characteristics of a lifting-body design, believing they could make conventional landings hazardous. The Air Force accepted the panel's recommendation, deciding to finance only winged reentry vehicle programs: the Boeing Aircraft Company's manned Dyna-Soar X-20 and McDonnell Aircraft Company's upiloted ASSET (Aerothermodynamic/elastic Structural Systems Environmental Test). Only a mock-up of the X-20 was ever built, Secretary of Defense Robert McNamara canceling the $458-million X-20 program in December 1963. In 1964, the $21-million unpiloted ASSET hypersonic glider was flown successfully four times in hypersonic reentry maneuvers. Never flown subsonically, the four ASSET research vehicles were parachuted into the ocean for recovery.[3]

Meanwhile, as NASA decision-makers continued to stay with ballistic shapes for the Mercury, Gemini, and Apollo programs, some NASA researchers at the field centers continued to study lifting-body reentry configurations. Actually, interest in the lifting-body concept among individuals at NASA dates back to the early 1950s when researchers—under the direction of two imaginative engineers, H. Julian "Harvey" Allen and Alfred Eggers—first developed the concept of lifting reentry from suborbital or orbital space flight at NACA's Ames Aeronautical Laboratory at Moffett Field in California. In March 1958, the researchers presented this work at a NACA Conference on High-Speed Aerodynamics.[4]

The initial work of NACA researchers in the early 1950s had been done in connection with studies regarding the reentry survival of ballistic-missile nose cones, the results of which were first reported in 1953. Researchers found that, by blunting the nose of a missile, reentry energy would more rapidly dissipate through the large shock wave, while a sharp-nosed missile would absorb more energy from skin friction in the form of heat. They concluded that the blunt-nosed vehicles were much more likely to survive reentry than the pointed-nose vehicles. Maxime A. Faget and the other authors of a paper at the 1958 NACA Conference on High-Speed Aerodynamics concluded that "the state of the art is sufficiently advanced so that it is possible to proceed confidently with a manned satellite project based upon the ballistic reentry type of vehicle." Faget's paper also indicated that the maximum deceleration loads would be on the order of 8.5g, or 8.5 times the normal pull of gravity on the vehicle.[5]

Other authors at the same conference presented the results of a study on a blunt 30-degree half-cone wingless reentry configuration, showing that the high-lift/high-drag configuration would have maximum deceleration loads on the order of only 2g and would accommodate aerodynamic controls. This configuration also would allow a lateral reentry path deviation of about plus or minus 230 miles and a longitudinal variation of about 700 miles.[6]

---

3. See Richard P. Hallion, "ASSET: Pioneer of Lifting Reentry," in Hallion, ***Hypersonic Revolution***, pp. 449-527.

4. National Advisory Committee for Aeronautics, ***NACA Conference on High-Speed Aerodynamics, A Compilation of the Papers Presented*** (Moffett Field, CA: Ames Aeronautical Laboratory, 1958). Notable in this connection was the paper by four Ames researchers—Thomas J. Wong, Charles A. Hermach, John O. Reller, Jr., and Bruce E. Tinling—at a session chaired by Allen. The paper's title was "Preliminary Studies of Manned Satellites—Wingless Configurations: Lifting Body" and appeared on pp. 35–44 of the volume just cited.

5. Maxime A. Faget, Benjamine J. Garland, and James J. Buglia, "Preliminary Studies of Manned Satellite Wingless Configuration: Nonlifting" in ***NACA Conference on High-Speed Aerodynamics***, pp. 19–33 with the quotation on p. 25.

6. Wong et al., "Preliminary Studies: Lifting Body," pp. 35–44.

Following this conference—the last held by the NACA before Congress created the NASA later in 1958—the logical choice for a piloted reentry configuration seemed to be the proposed blunt half-cone 2g vehicle with controls and path deviation capability rather than the 8.5g ballistic vehicle with no controls and almost no path deviation capability. However, this was not to be, due to some practical considerations of the time.

As things turned out, the thrust capability of the available boosters versus the needed payload weights made it easier to design a small blunt shape to fit on top of the Redstone and Atlas rocket boosters. This blunt-nosed ballistic configuration became the United States' first piloted spacecraft, the Atlas rocket-boosted Mercury capsule, which then evolved into the Apollo program using the Saturn rocket.

Nevertheless, the concept of wingless lifting reentry did not die. The only problem was that we had no experience with this type of vehicle, especially with the anticipated heat loads. But the advantages of a blunt half-cone or wingless reentry vehicle over the space capsules are easy to understand.

"Lifting" reentry is achieved by flying from space to a conventional horizontal landing, using a blunt half-cone body, a wingless body, or a vehicle with a delta planform (like the shape of the current Space Shuttle), taking advantage of any of these configurations' ability to generate body lift and, thus, fly. We could not put conventional straight or even swept wings on these vehicles because they would burn off during reentry —although a delta planform with a large leading-edge radius might work. These vehicles, or lifting bodies as we called them, would have significant glide capability down-range (the direction of their orbital tracks) and/or cross-range (the direction across their orbital tracks) due to the aerodynamic lift they could produce during reentry.

Space capsules, on the other hand, reenter the Earth's atmosphere on a ballistic trajectory and decelerate rapidly due to their high aerodynamic drag. In short, although capsules can produce small amounts of lift, they also generate large amounts of drag, or resistance. Space capsules are subject to high reentry forces due to rapid deceleration, and they have little or no maneuvering capability. Consequently, capsules must rely on parachute landings primarily along the orbital flight path.

In contrast, a lifting body's ability to produce lift and turn right or left from the orbit would allow any one of many possible landing sites within a large landing zone on both sides of the orbit on the return to earth. Furthermore, deceleration forces are significantly reduced with a lifting-body vehicle, from about 8g to 2g. The lifting-body landing "footprint" for a hypersonic vehicle—that is, one with a speed of Mach 5 or greater and a lift-to-drag ratio of 1.5—includes the entire western United States as well as a major portion of Mexico, a significant improvement over that of a capsule. The prospect of achieving these advantages of lifting reentry was rather exciting, given the limited capability of ballistic reentry capsules.

Free-Flight Model of the M2-F1 Lifting Body

Fascinated by the possibility of an airplane that flies without wings, I began talking in 1962 to other engineers and engineering leaders at the NASA Flight Research Center and at the NASA Ames and Langley research centers. I found skepticism to be abundant, many believing, as various design studies at the time had suggested, that some sort of deployable wings would be needed to make a lifting body practical for landing. Some of the most conservative design studies, not content to stop with deployable wings, even suggested that deployable turbojet engines should be used. Obviously, the space and weight allotment in these designs left little, if any, allowance for payloads. Even more design reports on lifting bodies gathered dust on library shelves as even more decisions were made to use symmetrical reentry capsules for spacecraft programs.

About this time, it occurred to me that for lifting bodies to be considered seriously for future spacecraft designs, some sort of flight demonstration would be needed to boost confidence among spacecraft designers regarding lifting bodies. At first, I limited myself to launching countless paper lifting-body gliders down the halls, while behind my back passersby sometimes rolled their eyes and made circling-finger-at-temple motions. Then, as much to satisfy my own growing curiosity as to demonstrate lifting-body flight potentials to my peers, I constructed a free-flight model in a half-cone design that was very similar to what would later become the M2-F1 configuration.

I made the frame with balsa stringers and the skin out of thin-sheeted balsa. Adjustable outboard elevons and adjustable vertical rudders made up the control system. I began with the center of gravity recommended in Eggers' design studies, then changed it with nose ballast. For landing gear, I used spring-wired tricycle wheels.

I hand-glided the model into tall grass as I worked out the needed control trim adjustments. The model showed characteristics of extremely high spiral stability. The effective dihedral (roll due to a side gust) was very high, and launching the model into a bank would cause it to roll immediately to the equivalent of a wings-level position.

Expanding the flight envelope, I then started hand-launching the model from the rooftops of buildings for longer flight times. The outer elevons were effective but not overly sensitive to adjustments for longitudinal trim and turning control. Experimenting with the vertical rudders, I found the roll response very sensitive to very small settings of the rudders. In these first flights of the model, I did not experiment with body flaps. The model had a steep gliding angle, but it would remain upright as it landed on its landing gear.

Next, I towed the model aloft by attaching a thread to the upper part of the nose gear, then running as one does in lifting a kite into flight. The model was exceptionally stable on tow by hand. Naturally, I then thought of towing the model aloft with a gas-powered model plane since I just happened to have a stable free-flight model that I had used successfully in the past to tow free-flight gliders.

Attaching the tow-line on top of the model's fuselage, just at the trailing edge of the wing, created minimum effect on the tow plane from the motions of the glider behind it. After sufficient altitude was reached for extended flight, a free-flight vacuum timer released the lifting-body glider from the tow-plane. All flights of the model were done at Pete Sterks' ranch east of Lancaster, an area where most of the NASA Flight Research Center employees lived at the time. From Sterks's ranch, I had also flown other model airplanes as well as my 65-horsepower Luscombe light plane.

I found the inherent stability of the M2-F1 lifting-body model, both in free flight and on tow, very exciting—so much so that I knew it was time to make a film to show my peers and bosses just how stable it was in flight. To film the flight of the M2-F1 lifting-body model, I enlisted the help of my wife, Donna, and our 8mm camera. We made the film on a nice, calm weekend morning at Sterks' ranch. While I prepared the tow-plane and the M2-F1 model for launch, Donna stretched out on the ground on her stomach to film the flight from a low angle, making the M2-F1 model look much larger than it actually was.

Both the flight and Donna's film-making were successful, the film showing the M2-F1 stable in high tow, then gliding down in a large circle after the timer released it from the tow-plane. The lifting-body model reached the ground much sooner than did the tow-plane because the lifting body's much lower lift-to-drag ratio gave it a much steeper gliding angle. The M2-F1 made a good landing on Sterks' dirt strip, while the tow-plane landed unharmed in the alfalfa field next to the landing strip. Since I was just getting started in radio control at the time, I used the free-flight approach in these early flights to keep things lightweight and simple. Later, I towed the M2-F1 lifting-body model with a radio-controlled tow-plane.

Starting a Lifting-Body Team

My mounting enthusiasm began to rub off on my peers at the NASA Flight Research Center. The first to join my lifting-body cause was a young research engineer named Dick Eldredge. (In fact, we were all young at the time.) A graduate of Mississippi State's aeronautical engineering department, Dick had been a student of an aerodynamicist named August Raspet, who had established a flight-test facility at a landing strip near the university where he involved many of his students, including Dick, in flight research. As a result, Dick had brought with him to the NASA Flight Research Center a great deal of skill and enthusiasm regarding the aerodynamics and structures of aircraft design.

Having built three gliders on his own, Dick had excellent skills in design and fabrication of structures in welded steel, wood, and aluminum sheet-metal. At the time, the NASA Flight Research Center also had a small "Skunk Works" second to none in its skilled machinists, aircraft welders, sheet-metal workers, and instrument builders. Dick knew each of these craftsmen personally, not just at work but much more through contact with them on the weekends, many of these NASA craftsmen also being involved with their own airplane-building home projects.

Dick Eldredge and Dale Reed resting their arms on the M2-F1. In the background is a Space Shuttle, which benefited from lifting-body research. (NASA photo EC81 16283)

Dick Eldredge and I made a strange but good team. Since I was tall and he was very short, some people thought of us as a "Mutt & Jeff" duo. Together, we would critique and challenge each other's ideas about how to solve design problems until we mutually came up with the best solutions. We never wasted time belaboring the problem but, after agreeing on a solution, went on to the next design challenge. I always thought of Dick as my "little buddy."

Dick and I enjoyed bouncing ideas off one another for new aircraft designs. At the time, the British Kramer Prize had not yet been awarded for the world's first man-powered airplane. Each year, the prize became more enticing to us as it grew in size to $100,000 and opened to persons beyond Britain throughout the world. At lunchtime, Dick and I plotted and schemed on how we could win the Kramer Prize. Dick had done a lot of research on the various British designs that, while they could fly in a straight line, could not make the required figure eight. Most of these designs included hundreds of parts and took hundreds of hours to build. Dick and I agreed that the winning design would have to have very low wing loading and be simple to build and repair. Unfortunately, we both were young enough to have growing families that required a great deal of our time at home, so Dick and I never had the time or means for an after-hours project of the sort that might win the Kramer Prize.

However, Dick and I had a mutual friend in Paul McCready of Pasadena, who, about the time we were forced to abandon our man-powered project, got his family involved in a similar project, helped along by a number of volunteers with skills in model-building and bicycle-racing. McCready put into action the low wing loading and simple structural approach that Dick and I had only been able to talk about.

At Taft, not too far from Edwards Air Force Base, McCready demonstrated the world's first man-powered flight with the Gossamer Condor. His first flight tests of the Gossamer Condor, in fact, had been at Mojave, just down the road from Edwards AFB. McCready went on to build a second craft called the Gossamer Albatross, which the bicyclist Bryan Allan piloted across the English Channel.[7]

Afterwards, I worked with McCready and a backup Gossamer Albatross on a flight research program at the NASA Flight Research Center, having gotten approval to

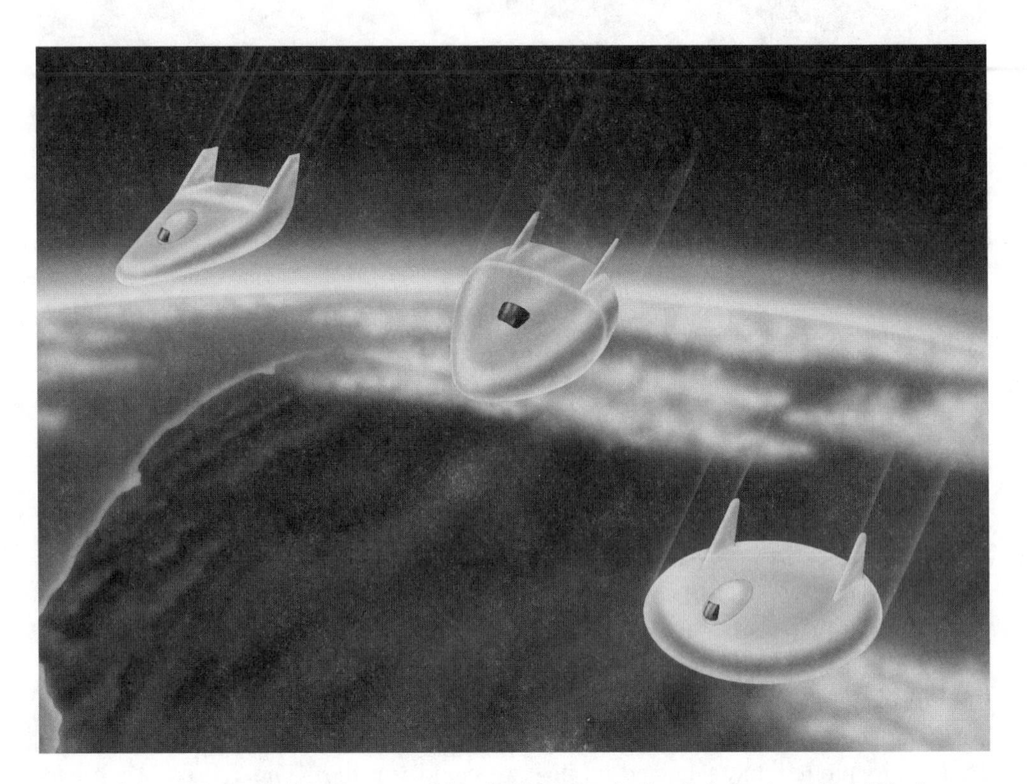

Proposed Ames M2-F1, M1-L half-cone, and Langley lenticular bodies. Dale Reed and Dick Eldredge proposed testing the three shapes using a common internal structure for all of them. (NASA photo E62 8933)

7. See M. Grosser, "Building the Gossamer Albatross," ***Technology Review*** 83 (Apr. 1981): 52–63; Paul McCready, "Crossing the Channel in the Gossamer Albatross," Society of Experimental Test Pilots, ***Technical Review*** 14 (1979): 232-43.

make this official NASA project to measure the aerodynamic characteristics of the aircraft with lightweight research instrumentation installed at the Flight Research Center. This successful program resulted in a published report on the aerodynamic characteristics of the Gossamer Albatross.[8]

As a team, once we both were bitten by the lifting-body bug, Dick and I developed a research plan for testing three lifting-body shapes with a common structural frame housing the pilot, landing gear, control system, and roll-over structure. The three lifting-body shapes were the Ames M2-F1, the M1-L half-cone, and the Langley lenticular.

The lenticular lifting-body shape was particularly intriguing because, to many of us, it immediately calls to mind the popular flying-saucer portrayed by the media as the spacecraft of extraterrestrials. My wife, Donna, however, had her own special appreciation of the lenticular shape, dubbing it the "Powder Puff."

All three of the lifting-body shapes were based on some sort of variable geometry. The M2-F1 was a 13-degree half-cone that achieved transonic stability by spreading its body flaps much like what's done by a shuttlecock in the game of badminton. The M1-L was a 40-degree half-cone that achieved a better landing lift-to-drag ratio by blowing up a rubber boat tail after it slowed down. The lenticular lifting-body would transition to horizontal flight by extending control surfaces after making reentry much like a symmetrical capsule.

Our concept was to construct the shapes separately, building three wooden or fiberglass shells that could attach to an inner structure common to all three shapes. If we could build the vehicles to be light enough, they could be towed by ground vehicles across the lakebed before being towed aloft by a propeller-driven tow-plane.

Dick suggested a control system that I liked instantly: a mechanical way of mixing controls that was similar to what is done now in modern high-tech aircraft by digital electronic control systems. The scheme was to connect a swashplate on the aft end of the steel-tube structure to the pilot's control stick and rudder pedals. The swashplate, pivoting on one universal joint, took up various positions, depending on the combination of roll, pitch, and yaw commands the pilot sent to the front side of the swashplate. With push rods hooked up to different locations on the backside of the swashplate, and to the horizontal and vertical control surfaces on the aft end of the lifting-body shapes, any combination in control-mixing could be achieved. These controls could be altered easily during the flight-test program or changed to fit another lifting-body shape.

---

8. Henry R. Jex and David G. Mitchell, "Stability and Control of the Gossamer Human-Powered Aircraft by Analysis and Flight Test" (Washington, D.C.: NASA Contract Report 3627, 1982).

Milt Thompson Joins the Lifting-Body Team

Fired by enthusiasm, Dick and I kept charging down the design road with the lifting-body research vehicles so that we could make a pitch to our boss, Paul Bikle, to gain his support for a lifting-body program. One day I told Dick, "You know, if we had a pilot on our team, we would have a much better chance of selling the program concept." Then, we talked to Milt Thompson, whom we saw as the NASA test pilot most likely to be interested in our project.

Milt was a skilled pilot with a distinguished background as a Naval aviator, Boeing flight-test pilot, and NASA research pilot. As one of the twelve NASA, Air Force, and Navy pilots who flew the X-15 between 1959 and 1968, Milt had fourteen flights in the rocket-powered aircraft to his credit, reaching on separate occasions a maximum speed of 3,723 mph and a peak altitude of 214,000 feet.

Earlier in 1962, before Dick and I talked to Milt about the lifting-body project, the Air Force had selected Milt to be the only civilian pilot for the X-20 Dyna-Soar program scheduled to launch a man into earth orbit and recover with a horizontal ground landing, a program later canceled shortly after construction had begun on the X-20 vehicle. Not having an ego problem, Milt loved flying unusual or unorthodox aircraft configurations as varied as the rocket-powered X-15 and the ungainly Paresev, a vehicle designed and built at the NASA Flight Research Center's "Skunk Works" to test the Rogallo Wing concept for spacecraft recovery.

Milt was easy to talk to and could relate readily to flight research engineers. Very methodical in planning flights, he did not take risks beyond the unavoidable ones normal for first-time aircraft configurations, a characteristic that earned him high regard from both pilots and project managers. A handsome, wild, and wonderful guy, Milt had a winning personality and persuasive charm. All the women seemed to be in love with him. Popular, he was a friend to everyone. Dick and I knew that Milt was the guy who could help us sell the lifting-body program.

We presented to Milt our idea for testing lifting bodies, asking him if he would join us and fly a lifting body—if and when we got one built. Without hesitation, he gave us a solid "yes." Now we were a team of three.

The three of us put our heads together to decide on the next step to take in promoting our program. Milt suggested that if we had one of the originators of the lifting-body reentry concept on our side, we could move our cause along more rapidly.

I phoned Al Eggers at the NASA Ames Research Center, located at Moffett Field in northern California, and described our idea to him. Very enthusiastic, Eggers asked how he could help. At the time, Eggers was a division head at Ames in charge of most of the wind tunnels. We were going to need a lot of support in wind-tunnel tests if we were to figure out how to fly these crazy aerodynamic shapes.

Telling Eggers that we were preparing a pitch to sell the idea to Paul Bikle, I asked him if he would like to hear the pitch. "Definitely," he replied. We arranged a meeting at the Flight Research Center so we could present our idea to both Paul Bikle and Al Eggers.

I presented a simple program plan for building the vehicles in the "Skunk Works" shops at the Flight Research Center, instrumenting the vehicles, and then flight-testing them to measure stability, control, and other aerodynamic characteristics in flight. After I presented the preliminary design drawings that Dick and I had made, I showed the film that my wife had made of my model M2-F1 flights.

Milt Thompson's endorsement of the plan pushed it over the crest. We received a hearty "yes" from both Bikle and Eggers. Eggers offered full use of the wind tunnels for getting any data needed to support the program if Bikle would be responsible for developing and flight-testing the lifting-body vehicles. It was agreed as well, however, that we would take it one step at a time, starting with the M2-F1 configuration and building it as a wind-tunnel model to be tested in the 40-by-80-foot wind tunnel at Ames.

Armed with a cause and fired with enthusiasm, we found ourselves gaining more and more support from our peers. We even came up with an unofficial motto for our lifting-body reentry vehicle project: "Don't be rescued from outer space—fly back in style." With the space program then dependent on the ballistic capsules, the astronauts were being fished out of the ocean, sometimes nearly drowning in the process and usually after some degree of sea sickness. Wen Painter—later a prominent engineer in the rocket-powered lifting-body program—drew a cartoon depicting the difference the lifting-body reentry vehicle would make in how astronauts would return to earth, his cartoon showing the astronaut landing at an airport in style, greeted by a reception hostess.

If the great enthusiasm of the builders at the NASA Flight Research Center resulted in a wind-tunnel model capable of actual flight, well, as Bikle noted, that would be something simply beyond the control of management. Later, we would go through the official process of getting approval from NASA Headquarters for flying the vehicle. In the meantime, we decided to get to work while everyone was enthusiastic and ready to start. With this decision, the lifting-body program was launched.

CHAPTER 2

"FLYING BATHTUB"

Our goal was to design and build a very lightweight vehicle that could be towed across the lakebed with a ground vehicle and, later, aloft with a light plane, the way sailplanes are towed. Based on the tiny model used in the filmed flights, the first lifting-body vehicle was also called the M2-F1—the "M" signifying a manned vehicle and the "F" designating flight version, in this case the first flight version.

Months before the M2-F1 was completed, it had already been dubbed the "flying bathtub" by the media. The first time seems to have been on 12 November 1962 in the ***Los Angeles Times*** article "'Flying Bathtub' May Aid Astronaut Re-entry." The article included a photo of Milt Thompson sitting in a mock-up of the M2-F1 that, indeed, looked very much like a bathtub.

Paul Bikle decided to run the project locally, financing it entirely from discretionary funds. He thought that a volunteer team at the NASA Flight Research Center, supplemented with local help as needed, could build the M2-F1 faster and cheaper than NASA Headquarters could through a major aircraft company. As history proves, Bikle was right.

The M2-F1 was built entirely in four months. Engineers at the Flight Research Center also kept the cost of designing, fabricating, and supporting the M2-F1 to under $30,000, about the cost of a Cessna. At the time the M2-F1 was built, someone associated with a major aircraft company was cited anonymously as saying that it would have cost an aircraft company $150,000 to build the M2-F1. The extremely low-cost M2-F1 program would have invaluable results later, proving to be the key unlocking the door to further lifting-body programs.[1]

A Matter of Teamwork: Building the M2–F1, 1962-1963

After our meeting with Paul Bikle and Al Eggers, we were swiftly swept up into the enthusiastic atmosphere of the lifting-body program. On his return to the NASA Ames Research Center, Eggers asked Clarence Syvertson, his deputy, to coordinate all wind-tunnel tests that we needed in support of our design and flight-planning activities. Meanwhile, at the NASA Flight Research Center, Bikle asked me to put together a team to design and fabricate the first lifting-body vehicle.

Long before I began to put together that team, Dick Eldredge and I had already fairly much agreed that the basic design would include two structural elements, a core

---

1. Stephan Wilkinson, "The Legacy of the Lifting Body," ***Air & Space*** (April/May 1991), p. 51; Hallion, ***On the Frontier***, p. 149.

M2-F1 fin fabrication by Grierson Hamilton, Bob Green, and Ed Browne. (NASA photo E94 42509-13)

steel-tube structure and a detachable aerodynamic shell. However, the real work lay ahead of us in the detailed design of the hundreds of parts needed for the actual vehicle.

To do this work, I selected a design-and-fabrication team made up of four engineers and four fabricators, all of whom were aircraft buffs involved with home-building their own airplanes, most of them members of the Experimental Aircraft Association. These individuals had worked together to some extent on previous programs in the Flight Research Center's unofficial "Skunk Works." The group's chief designer was Dick Eldredge. To lead the team, we got Vic Horton, a no-nonsense operations engineer who took pride in keeping to schedules.

Horton picked up a few extra part-time volunteers as the work got underway. Hardware designers, besides Eldredge, included Dick Klein and John Orahood. Meryl DeGeer calculated stress levels in the structure to verify the adequacy of the design. Ed Browne, Howard Curtis, Bob Green, Grierson Hamilton, Charles Linn, George Nichols, and Billy Shuler fabricated the internal steel-tube carriage of the M2-F1 as well as its aluminum sheet-metal tail fins and controls.

Once we had the initial team, we needed a place to work. We sectioned off a corner of the fabrication shop with a canvas curtain, labeling it the "Wright Bicycle Shop." Indeed, we felt very much like the Wright Brothers in those days, working at the very edge between the known and unknown in flight innovation. In the "Wright Bicycle Shop," we put the drafting boards next to the machine tools for maximum communication between designers and fabricators. This strategy worked extremely well. A fabricator could lean over a designer's drafting board and say, "I could make this part faster and easier if you would change it to look like this."

I think our project was Bikle's favorite at the time. We would see him at least once a day, and we got a great deal of extra attention from him. A few chose at the time to grumble about Bikle acting as if he were the super project engineer on the M2-F1, but I think that we thrived as a team from his presence. For one thing, I have never since seen on later projects enthusiasm or morale as high among team members as existed on the M2-F1 project. In fact, it really isn't an exaggeration to say that we had trouble keeping the team from working through lunch, during the evenings, or on weekends.

The M2-F1 project also benefited from Bikle's experience and suggestions. While we didn't have to use his suggestions, we did need to have good reasons for not using them. A time when one of his suggestions helped us a great deal—and there were many such times—was when we had everything else thought out and had begun trying to decide how to build the aerodynamic shell.

The core of the dilemma had to do with the shell's weight, which we hoped to keep under 300 pounds, wanting a vehicle of minimum weight so that the M2-F1 would fly slowly enough that a ground vehicle could tow it aloft. Dick Eldredge and I were thinking about building the shell out of fiberglass, but we weren't sure we could keep the weight within necessary limits.

We knew that our vehicle design lent itself easily to being built in two different locations by two different teams, the two main assemblies being joined later. We knew that we could build the internal steel-tube carriage, tail surfaces, and controls in our NASA shop while the outer shell was being built elsewhere by a second team. But where and by whom?

Bikle suggested that we talk to a sailplane builder named Gus Briegleb, who operated an airport for gliders and sailplanes at El Mirage dry lake, 45 miles southeast of Edwards Air Force Base. Bikle also suggested that he might be able to find enough money in his discretionary fund to contract Briegleb to build the shell for us out of wood.

One of the nation's last artisans building aircraft out of wood, Briegleb had founded the Briegleb Glider Manufacturing Company during World War II to design and build wooden two-place trainer gliders for Army pilots being trained to fly troop-assault gliders. The two-place trainer gliders were used to train these pilots to fly in formation on a tow-line and performing precision dead-stick landings after release from Navy R4D tow-planes (same as the Air Force C-47). The troop gliders were used extensively during the Allied invasion of France, with the Briegleb Glider

Manufacturing Company being one of only a few companies manufacturing the trainers.

In 1962, when we contacted him, Gus Briegleb was trying to keep alive the art of fabricating wooden airplanes by selling kits of a high-performance sailplane that he had designed. Between selling these kits and operating the glider-sailplane airport at El Mirage, Gus was making a living, but he definitely was not getting rich.

Briegleb responded enthusiastically when we approached him about building the M2-F1 shell out of wood. Although wood eventually gave way to aluminum sheet-metal in the production of aircraft for a good number of excellent reasons, wood is still one of the more efficient structural materials for aircraft in terms of fatigue life, vibration damping, and strength-to-weight ratios. Briegleb initially proposed to build the shell out of wood for only $5,000.

Thinking that sum was too low, Bikle asked Briegleb if he had considered overhead, profit, and unforeseen problems that were likely to arise during the building of the shell. A builder, not a businessman, Briegleb admitted he had not considered these things. Bikle said that he could authorize up to $10,000 for the wooden shell, that being at the time the limit for small purchases at the NASA Flight Research Center. Briegleb agreed to meet the 300-pound target weight and the strength specifications that Dick Eldredge and I had determined from airload calculations, and he agreed to deliver the shell four months from the date the contract was signed.

Wooden shell of M2-F1 at El Mirage, seen from the rear. (NASA photo E94 42509-10)

When Briegleb got into the detailed design process, he found that the shell would have to be far more complicated than he had originally thought to keep it to the specified weight. He had underestimated the hours needed to build the shell by at least a factor of three. The shell had to be made with two internal keels to carry the loads to the steel-tube frame. Hundreds of small wooden parts made up these built-up wooden keels. To support the outer skin shape, the keels also had multiple internal cross-bracings made of miniature wooden box beams of webs and spar caps, all nailed and glued together.

When we saw the predicament that Briegleb was in, we sent him some help: Ernie Lowder, a NASA craftsman who had worked on building Howard Hughes' mammoth wooden flying-boat, the Spruce Goose. Despite having Lowder as a full-time fabricator, Briegleb says he still ate quite a bit of the $10,000 contract. Nevertheless, Briegleb was very proud of his work, and so were we. He delivered the shell to us on time, at cost, and slightly under the 300-pound weight limit. I think we gained a great advantage by being able to use the last of America's finest wooden-airplane craftsmen to build the shell of the M2-F1.

As Briegleb's team built the outer shell, NASA craftsmen built the internal steel-tube structure. The steel-tube carriage was finished first, in about three months, and while the wooden shell was still being fabricated at El Mirage, the carriage was being rolled around on the landing gear. Eldredge and I had designed the M2-F1 so that it took only four bolts to attach Briegleb's shell to the internal structure.

Team Three for Analysis

Once the two teams were in place and building the two main structures of the M2-F1, I realized we also needed a third team to do the analysis on aerodynamics, control rigging, and characteristics of stability and control to support flight tests. Using the wind-tunnel data on small-scale lifting-body models that we were beginning to get from the NASA Ames Research Center, I could determine the basic stick-to-surface gearing in pitch for the outer elevon surfaces and the upper body flap. Rotating the lifting body nose-up to moderate angles of attack amplified to high angles the flow on the aft sides of the bulbous M2-F1 shape.

Tufts of yarn on the small-scale wind-tunnel model had indicated that its outer elevon surfaces experienced about twice the change in angle of attack experienced by the model's nose. Consequently, I specified gearing for the outer elevons to move three times more than the body flap with fore and aft travel of the pilot's control stick. I did this so that, when a differential roll side input was made from the pilot's stick, there would be no risk of stalling an elevon surface, causing reversal of the roll or loss of control of the vehicle during the roll.

Determining control rigging and gearing for turn control was not as obvious as that for pitch control. The M2-F1, and almost all of the later lifting bodies, have extremely high dihedral—that is, with wind from the side (called "sideslip"), the vehicle wants to roll in the opposite direction. Because of this characteristic, rudder deflections actually resulted in roll rates higher than those produced by differential elevon deflections. Since lifting bodies also have extremely low roll damping from having no wings to resist roll rates, and since Dutch roll results from the extremely high dihedral inherent in most lifting bodies, we had a potentially dynamic problem in stability and control if we did not do the right thing in designing the control system.

Obviously, we needed help from the experts in stability and control at the NASA Flight Research Center, all of whom were currently working on the X-15 program. In its later stages after three years, the X-15 program still had number-one priority at the Center. Because the X-15 program was so well organized and ran so smoothly by that time, many aspects were getting to be routine, even though there were still some surprises showing up during the speed and altitude buildup as the flight envelope was being expanded. Our unofficial lifting-body project was able to recruit the help it needed, despite the on-going X-15 program, thanks to Bikle's policy that the NASA Flight Research Center had an equal responsibility to aeronautical research directed to the future.

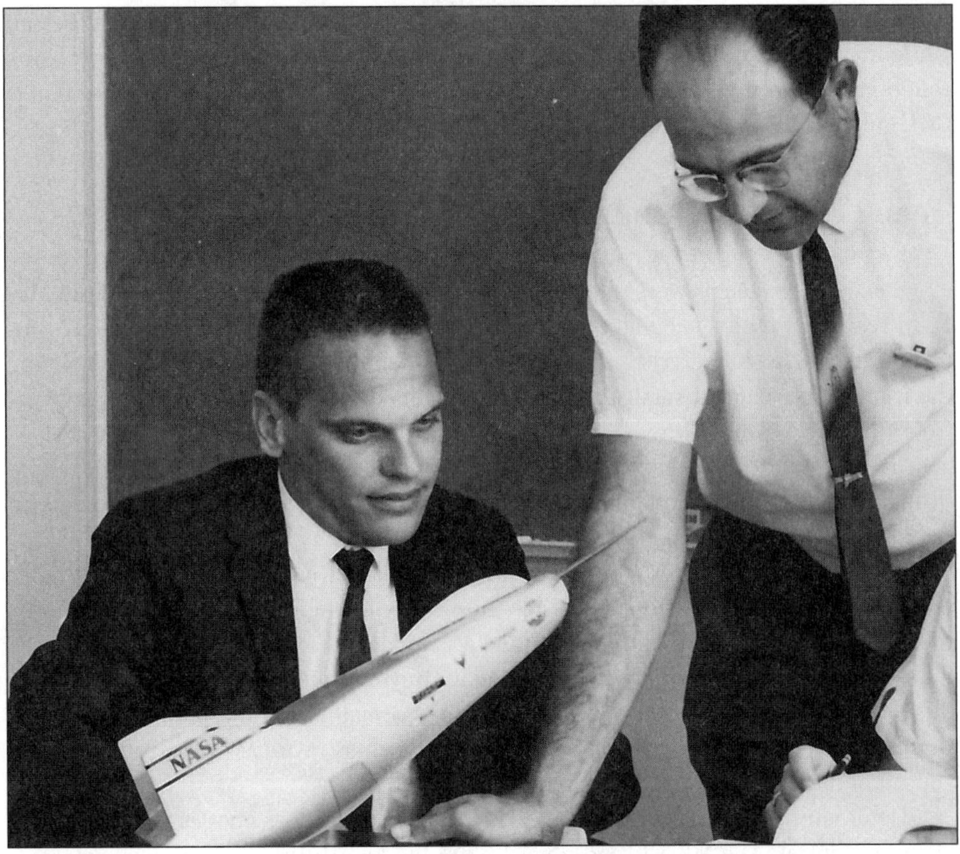

Ken Iliff, first member of the lifting-body analytical team, with Dale Reed. (NASA photo E66 15469)

My first volunteer was Ken Iliff, now the Chief Scientist at NASA Dryden, who at the time was a bright and enthusiastic twenty-one-year-old engineer just out of college and doing a mundane analytical task in reducing X-15 flight data. Iliff poked his head in the office where Eldredge and I were working and, after inquiring what we were doing, asked if there was anything he could do to help us out. "Sure!" I replied quickly, not one to refuse any help I could get.

After explaining to Iliff that we planned to get a high-speed ground-tow vehicle to tow the full-scale M2-F1 model across the dry lakebed, I asked him to take a stab at calculating what the rotation and lift-off speeds would be on ground-tow, information we needed in determining the requirements for the tow vehicle. We could have a problem, I explained, if the aerodynamic pitch controls were not strong enough to lift the nose, overcoming the nose moments from the wheel drag and the tow-line force.

Iliff got busy. He calculated rotation speed to be 59 miles per hour and lift-off speed to be 85 miles per hour. Later, when we actually ground-towed the M2-F1, we measured rotation speed at 60 miles per hour and lift-off speed at 86 miles per hour. Needless to say, we were impressed with this young engineer.

Mathematical Voodoo

Although he continued to maintain his obligations to the X-15 program, Iliff became more and more involved in our little lifting-body program. He started looking at the stability and control characteristics of our strange bird—just in case we did try to fly the M2-F1 following the full-scale wind-tunnel tests. Iliff sought help from his mentor, Larry Taylor, another engineer then studying pilot-control problems on the X-15 who was experienced in applying some of the latest techniques in analyzing stability and control problems on new aircraft configurations. Although Taylor had applied some of those techniques to the X-15 with success and gained the credibility of a number of his aerospace peers, some of the old-time flight-test engineers, including Paul Bikle, considered Taylor a radical practicing a kind of engineering witchcraft.

Taylor claimed he could use mathematics to describe the piloting characteristic of a test pilot, then predict the outcome of a planned flight. He called this the "human transfer function." Bikle disagreed, saying there was no way to predict how a pilot would perform on any one day, emphasizing that a pilot's performance was impacted by events in his personal life, such as having a spat with his wife or partying the night before a flight.

I felt both viewpoints had validity. I agreed with Taylor's viewpoint that there are fundamental differences in how individual pilots react to a difficult control task. In a stressful situation that leads to problems with pilot-induced oscillation, the gains of some pilots rise much faster than those of other pilots. An aircraft can go out of control if it has a tendency to oscillate in a particular direction, especially if the pilot tries to stop the oscillation by chasing the aircraft with the controls. Sometimes the airplane will halt the oscillations on its own if the pilot will slow down or stop moving the controls. However, this is not the usual or most natural reaction for a pilot during a stressful situation, for as arm and leg muscles tighten up from stress, control movements usually increase.

The master sorcerer of mathematical voodoo, Larry Taylor, was at the time passing his mystical art on to his apprentice, Ken Iliff, especially a strange engineering plot called "root locus" that many pilots then thought was pretty far-out stuff. The three categorical ingredients of this mathematical potion were the airplane's aerodynamics, inertial data composed of weights from all parts of the airplane, and flight conditions such as speed, altitude, and angle of attack. The root-locus plot gave results for different types of pilots, ranging from the totally relaxed pilot who does nothing with the controls to the high-gain pilot who moves the controls rapidly.

One magical point on the plot called the pole represented the "do-nothing" pilot. Another magical point called the zero represented the high-gain pilot or autopilot. A line connected these two points, representing all pilots between the two extremes. If the line moved into the right side of the plot, the pilot/aircraft combination was deemed unstable, predicting loss of control of the aircraft. Despite the fact that in the early 1960s even a number of engineers considered the root-locus analysis to be some sort of witchcraft, today root locus is a common mathematical tool used by stability and control engineers.

According to Bob Kempel—then with the Air Force and later a stability and control engineer at the NASA Flight Research Center with considerable influence on the design of control systems for experimental piloted and unpiloted NASA aircraft—root locus is a tool by which engineers can predict potential instability prior to flight so that a possibly catastrophic situation can be avoided by either pilot training or modification of the flight control system. "The intent of the engineer," says Kempel, currently active in control-system designs, "is to provide the test pilot with a pilot/airplane combination that will remain stable, regardless of pilot gain," workload variations, or emergency control situations.

Well tutored by Taylor in this technique, Iliff set off to predict the M2-F1's qualities during flight. He modeled the lifting body mathematically for free-flight as well as for flight while on tow. He found that the tow-line force was quite high in opposing the high drag of the lifting body, adding a high level of static stability to the system, much like towing a high-drag target behind an aircraft.

About this time, two more volunteers showed up whose help would be invaluable on the M2-F1. Bertha Ryan and Harriet Smith, two junior engineers who did not have strong obligations to the X-15 program, asked me what they could do to help. In getting Ryan and Smith as well as so many other volunteers, I was enjoying a bit of luck. The 50 percent of the work force at the NASA Flight Research Center not committed to the X-15 program wasn't being taxed fully in support of other official NASA programs. Even in those days, bureaucratic methods of operation caused tremendous lags to appear between approval and funding cycles. Furthermore, peaks and valleys in workloads occurred at the field stations whenever NASA Headquarters approved, turned down, or canceled a program, no matter how well the field managers scheduled work.

I was one of those Johnny-on-the-spot opportunists who would move in with my small program to take advantage when valleys appeared in workloads. Most supervisors liked to keep their people busy, and it didn't hurt the lifting-body project one bit to have the local director interested enough in our project to send us new volunteers. Bikle had encouraged Ryan to work with us, knowing that since she owned her own sailplane, she would have practical as well as analytical skills useful to the project.

Although engineers today are as often women as they are men, women engineers were not common in the early 1960s. After they volunteered on the M2-F1 project, I explained to Ryan and Smith that Milt Thompson wanted some sort of simulator for practice before flying the M2-F1. Good friends, Ryan and Smith thought the task would be fun. They also liked the idea of working as an all-woman simulation team—perhaps one of the first for those times—with Ryan preparing the aerodynamic data input and Smith mechanizing the simulation. Neither of them had ever set up a flight simulator before, but they felt that while the task would be challenging, they could also learn quite a bit by doing it. Actually, all of us were fairly naive about simulators in those days, even though a simulator had been set up for the X-15.

Harriet Smith, a member of the lifting-body simulation team and also of the analytical team. (NASA photo E58 3731)

Bertha Ryan, another member of the lifting-body simulation team and also of the analytical team. (Private photo furnished by Bertha Ryan, NASA photo EC97 44183-2)

When Eldredge and I had designed the M2-F1 control system to be flexible, we had thought we were being clever, never realizing that we created a veritable Pandora's box. Instead of having just one version of the M2-F1 to set up on the simulator, we had as many as five, one for each way our variable control system could be hooked up in its swashplate design.

We had six control surfaces on the rear of the vehicle that we could hook up in any combination to the pilot's stick and rudder pedals for pitch, roll, and yaw control—two vertical rudders, two outboard elevons that we called "elephant ears," and two horizontal body flaps at aft top. We also had a removable center fin, but no lower body flaps.

Most of the simulators used at that time were purely analog, requiring 30 or 40 hand-adjusted electrical potentiometers (called "pots") to be set up for each simulation session. It was very easy to make a mistake while setting up these pots, especially by setting a switch to give a minus instead of a plus sign, or vice versa. The only way to guarantee a correct simulation was to require a verification process for each simulation session. Despite their inexperience in setting up a simulation, Ryan and Smith were very methodical. They kept good notes and records, working hard at doing a good job.

Since pitch control did not seem to be a problem on the simulator, we spent much of our time trying to determine the best way to control roll and yaw on the M2-F1. Early on, we decided to eliminate the center fin as well as the differential control on the body flap. The center fin only made the already high dihedral even higher. Besides, we already knew from small-scale wind-tunnel tests that we had plenty of directional stability from the two vertical side fins. By making the body flap single-pitch rather than split, it could be used like an elevator, eliminating the need for the center fin as a fence against adverse yaw from a body-flap elevon system. The shop team members had already fabricated a two body-flap system, but by the simple expedient of bolting the right and left flaps together, they made one large flap.

We had narrowed the lateral-directional control system down to two basic possible schemes. In the first control scheme, right stick deflection would move the outer elevons for roll to the right, and the right rudder pedal would move both vertical rudders to the right. In the second control scheme, right stick deflection would move both vertical rudders to the right, and the right rudder pedal would move the outer elevons for roll to the right. Working with us as a part of the analytical team by flying the simulator in the ground cockpit, Milt would give us a pilot's rating for each of the configurations we investigated. His rating system was on a scale of one to ten, depending on the difficulty of changing and holding headings.

Eldredge and I had fairly much made up our minds in favor of the first control scheme, intuition having told us that elevons or ailerons should be controlled by the stick while rudders should be controlled by the rudder pedals. We were shocked when Milt told us that he preferred the second control system. His reasoning was that roll rates resulting from the rudders being deflected were twice as high as those resulting from differential elevon deflection. Milt felt that he could control the vehicle by using

M2-F1 simulator cockpit. (NASA photo E63 10278)

proper piloting technique, and he said he would rather have the higher roll rates available to him if he needed them.

If any research pilot could use proper piloting technique, it was Milt Thompson. He was a cool, disciplined pilot who could think well during emergencies or under other stressful conditions. He had already proven several times during the X-15 program that he could and would work closely with engineers in solving potential flight problems. He also liked to understand fully the idiosyncrasies of an aircraft before he flew it. In my opinion, Milt Thompson belongs up there with Chuck Yeager in any estimate of historic greatness for test pilots. Milt Thompson not only had the same stick-and-rudder skill and coolness under fire that Chuck Yeager had, but he also had a certain elegance in thinking when dealing with engineers. Milt had such an air of modest dignity and credibility about him—what today might be called "charisma"—that when he said he preferred the second control system for the M2-F1, we listened to him, even though we didn't necessarily like his choice.

At this point, Iliff did a root-locus plot for both control systems. He determined that there was no problem involved with using the first control system, with its use of the elevons for roll control. However, he found there could be a large problem with the second system which used the rudders for roll control. With the second system—the one Milt preferred—the M2-F1 could be driven unstable in Dutch Roll, resulting in loss of control of the vehicle, if the pilot's gains were too high. Although Taylor was doing a good job in verifying the root-locus technique on the X-15 program, it was still too new to be accepted by others as a valid design or planning tool. Despite Iliff's conclusions, Milt still insisted on using the second control system. His plan for the first car-tow tests was to gently rotate the M2-F1 nose-up until it was flying a few inches off the lakebed before he made any rudder or control-stick inputs. Then, he would move the controls very slowly to test them out. If things didn't look good, he would set the vehicle back down on its wheels, and we could try the other control system.

While the simulator is a wonderful tool in designing aircraft and planning flights, simulator results must be interpreted very carefully. A heavy smoker, Milt would sit in the simulator's cockpit totally relaxed, a cigarette in one hand, flying with the other hand. Under those conditions, unlike those of actual flight, he had no tendency toward driving the Dutch Roll mode unstable, as Iliff had predicted he would in actual flight.

During the month between the completion of the internal structure and the completion of the wooden shell, Vic Horton decided to test the ground stability and control of the internal structure with landing gear. The wheels and nose gear assembly were taken from a Cessna 150 light aircraft. The pilot steered by foot pedals through the nose gear. Milt being away on a trip, X-15 research pilot Bill Dana volunteered to sit in the pilot's seat while the structure was towed by automobile across the dry lakebed.

Dana was soon having a great time, sashaying back and forth like a water skier at thirty miles an hour on a 300-foot tow-line behind the automobile. Having good control of the steering, Dana was building a lot of confidence. Then, he pulled far over to one side and pulled the tow-release to test it out. Unfortunately, he had been holding a large amount of rudder pedal to compensate for the side pull of the tow-line.

Suddenly, the vehicle veered sharply and started to roll over. Dana countered with the rudder pedal. A wild oscillation began, the M2-F1 steel-tube skeleton doing a wheely to the right, then a wheely to the left. Finally, Dana lost control, and the M2-F1 flipped over. Fortunately, the fabricators had built a strong rollover structure, and both pilot and vehicle came out of the episode without injury. Dana was embarrassed by the incident, and we kidded him mercilessly for years, saying we'd call on him again if we ever needed to run a manned structural test.

Final assembly of the M2-F1 began when the wooden shell arrived from El Mirage. We lowered the steel-tube internal structure, minus the landing gear, through a large rectangular cutout in the top of the wooden shell. We inserted the landing gear legs through holes in the shell and bolted them to the inner steel structure. Four bolts on the two wooden keels attached the shell to the inner steel structure. The aluminum tail surfaces, built in the NASA Flight Research Center shop, were then bolted onto the wooden shell, and controls were hooked up by push-pull rods. Finally, we attached to the shell a Plexiglas canopy, made by Ed Mingelle of Palmdale for the M2-F1 after Bikle recommended that we go to him since he was a specialist in making custom canopies for sailplanes. Exactly four months from the day when Bikle had told me to begin building, the completed M2-F1 rolled out of the "Wright Bicycle Shop."

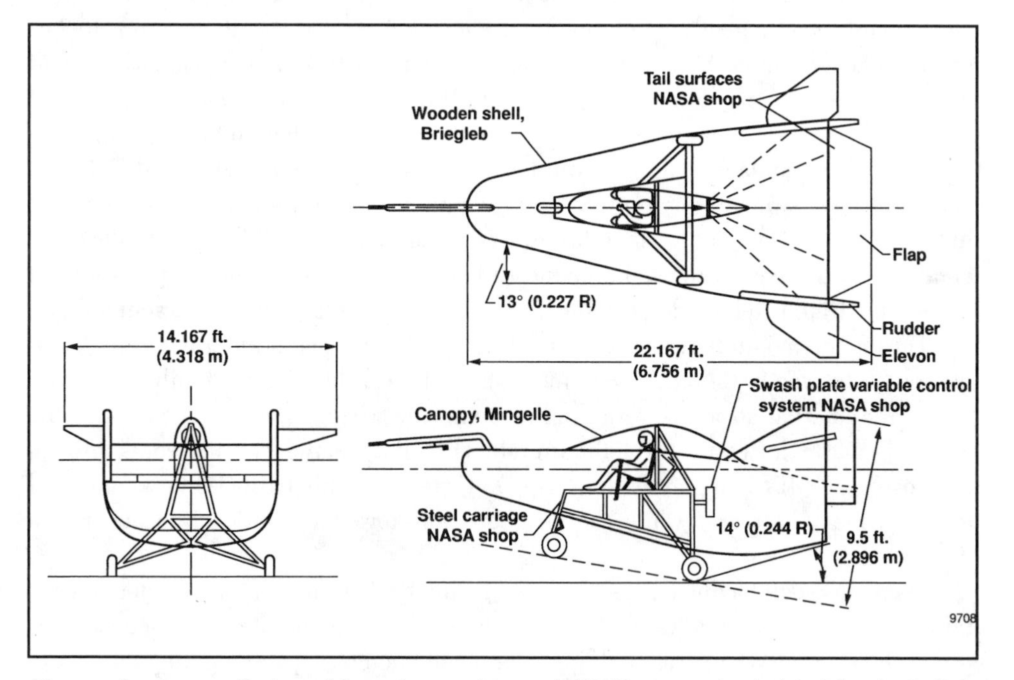

Diagram showing contributions of the various participants in M2-F1 construction (original drawing by Dale Reed, digital version by Dryden Graphics Office).

M2-F1 steel tube carriage that was tested by Bill Dana in car-tow tests prior to the installation of the M2-F1 shell. (NASA photo E63 10756)

NASA's Muscle Car: Ground-Towing the M2-F1

Dick Eldredge and I had designed the M2-F1 to weigh 600 pounds. However, like most prototype airplanes, it had grown in weight during fabrication, the completed vehicle weighing in at 1,000 pounds. From Iliff's calculations of the M2-F1's tow force and lift-off speed, we knew that to do taxi tests with the M2-F1 before the wind-tunnel tests at NASA Ames, we needed a ground-tow vehicle with greater power and speed than any of NASA's trucks and vans could provide.[2]

First, we needed a ground vehicle that could tow the M2-F1 at a minimum of 100 miles per hour. Secondly, we also needed a ground vehicle that, at that speed, could handle the 400-pound pull needed to keep the 1,000-pound lifting body airborne. In meeting these needs, we ended up with what was probably the first and only government-owned hot-rod convertible.

Once again, a volunteer came along who had the know-how that we needed. Working in operations at the NASA Flight Research Center at the time was Walter

2. Hallion, ***On the Frontier***, p. 150.

Pontiac tow vehicle next to the M2-F1. Walter "Whitey" Whiteside purchased the Pontiac by special order and had it modified in a hot-rod shop near Long Beach for its special mission. (Private photo furnished by Bertha Ryan, NASA digital image ED96 43663-1)

"Whitey" Whiteside, a retired Air Force maintenance officer who was also a veteran dirt-bike rider and expert hot-rodder.[3] Whitey volunteered to help us out by finding, purchasing, modifying, testing, maintaining, and driving the high-powered ground-tow vehicle that we needed.

At the time, the Pontiac Catalina seemed the best choice, this model having been the big winner the year before in Utah at the Bonneville Salt Flats time trials. With Boyden "Bud" Bearce's help in the procurement department, Whitey was able to make a special order from the factory for a Pontiac Catalina ragtop convertible with the largest engine then available, a four-barrel carburetor, and four-speed stick shift. NASA engineers at the Flight Research Center equipped the Pontiac with its tow rig and airspeed measuring equipment.

Whitey took the car for modification to Bill Straup's renowned hot-rod shop near Long Beach, where the straight-piped Pontiac was modified to run a consistent 140 miles per hour. There, auto-shop technicians also applied their hot-rod wizardry to the Pontiac, producing maximum torque at 100 miles per hour as measured by a

3. Milton O. Thompson, ***At the Edge of Space: The X-15 Flight Program*** (Washington, D.C.: Smithsonian Institution Press, 1992), p. 52.

dynamometer. They added a special gearbox, with transmission gear ratios significantly different from those that had helped the Catalina win at the Salt Flats, enabling the Pontiac eventually (once drag slicks were installed) to tow the 1,000-pound M2-F1 to 110 miles per hour in 30 seconds. The Pontiac's souped-up engine got about four miles to the gallon. Whitey got full support from the NASA fabrication shops headed by Ralph "Sparky" Sparks. Sparky and his right-hand man, Emmet Hamilton, took responsibility for keeping the Pontiac running and making any modifications required by Whitey.

For the safety of the driver and two onboard observers, Whitey had roll bars added to the NASA muscle car. He also had radios and intercoms installed. The front passenger bucket seat was reversed and the back seat was removed, replaced by another bucket seat so that a second observer could sit facing sideways. Of course, the Pontiac had to have government plates and the NASA logo on both sides. And just so no one would be encouraged to think the car was someone's personal toy paid for with government funds, the hood and trunk of the Pontiac were spray-painted high-visibility yellow so that the convertible looked just like any other flight-line vehicle.[4]

When the car was finished at the hot-rod shop, Whitey drove it back to the NASA Flight Research Center. A motorcycle fanatic and hot-rodder who loved speed, Whitey found it difficult to hold back once he got the Pontiac outside Los Angeles and on the highway across the desert. Realizing he would get his chance later to open up on the dry lakebed, he was being particularly careful to hold the Pontiac's speed to the posted speed limit when he saw in the rearview mirror the red light of a California Highway Patrol (CHiP) vehicle closely tailing the Pontiac. Pulling over to the side of the highway, Whitey wondered what he'd done wrong. It turned out that the officer was merely curious, having never before seen a government-owned convertible, especially one with a souped-up engine. After a careful up-close look and Whitey's explanation of how the car would be used, the officer drove away, shaking his head in amazement.

The Pontiac also caught the eye of other drivers whenever Whitey took it out onto little-traveled desert highways northeast of Edwards AFB through Four Corners, often into Nevada with its then anything-goes speed limits, to calibrate the car's speedometer, as typically done with research airplanes. Laughing, Whitey recently recalled one particular time when he headed out on just such a venture with one of the base's pilots

---

4. For other details on the Pontiac and its modification, see Hallion, ***On the Frontier***, pp. 150–151; Wilkinson, "Legacy of the Lifting Body," p. 54. For some details about ordering the Pontiac from the factory, intvw., Walter Whiteside by Darlene Lister, 21–22 June 1996. Both Hallion and Wilkinson identify the source of the modifications to the Pontiac as Mickey Thompson's shop. However, in an interview with Robert G. Hoey and Betty J. Love on 22 July 1994, Walter Whiteside was adamant that the source was Bill Straup's shop and that Mickey Thompson was the source for only the wheels and tires. This last interview is the source for several of the details in the present narrative.

in the car. As the Pontiac rumbled along, engine-exhaust system roaring as the speedometer moved above 100 miles per hour, Whitey glanced at the silent pilot, only to find him ashen-faced and trying to disappear into the seat.[5]

When we had the M2-F1 completed and ready for wind-tunnel testing at NASA Ames, we were still divided on which basic roll-control scheme to use. Bertha Ryan, Harriet Smith, and Milt Thompson backed their interpretation of simulation results, saying the rudders would give the best roll control. Ken Iliff and Larry Taylor countered with what their root-locus plots showed—that using the rudders for roll control would lead to pilot-induced oscillation. On the other hand, I thought that the outboard elevon surfaces simply looked right for roll control, and I believed that rudders were meant for yaw—not roll—control. In the end, we agreed to use the scheme Milt Thompson preferred, with the pilot's stick hooked to the rudders for roll control, as long as we could reconfigure to the other scheme if that one didn't work.

Of course, we had no official approval to flight-test the M2-F1, which was supposed to be merely a full-scale wind-tunnel model. Sitting in the cockpit, Milt Thompson reasoned that perhaps it wouldn't really be flying if we just lifted it off the lakebed a couple of inches. Boosting our confidence was the data we had from the earlier small-scale wind-tunnel tests. When approached, Bikle said to go for it, but to be careful.

We were very careful as we began on 1 March 1963, making several runs in car-tow at lower speeds, gradually working up to the nose lift-off speed of 60 miles per hour on 5 April 1963. During these runs, Milt became familiar with the cockpit and with visibility out the top, through the nose window at his feet, and out the side window level with his feet, these windows necessitated by the anticipated high angle of attack. He also became adept at nose-gear steering and using the differential brakes and tow-line release.

After a week of these cautious towings at lower speeds, Milt said he was ready to try a lift-off. Following Milt's radioed directions, Whitey took the Pontiac and the M2-F1 on tow up to 86 miles per hour, the 1,000-foot tow-line giving Milt plenty of maneuvering room.

Slowly Milt brought the nose of the little lifting body up until the M2-F1 got light on its wheels. Then, something totally unexpected happened. The M2-F1 began bouncing back and forth from right to left. Milt stopped the bounce by lowering the nose, putting weight back on the wheels. Several times he again brought the nose up until the M2-F1 was light on its wheels, and each time the vehicle reacted the same way, Milt ending the bounce by lowering the nose as he had the first time.

Later, in our little debriefing room, Milt said that he felt that if he had lifted the M2-F1 off its wheels, it would have flipped upside down in a roll. We started theorizing about the cause of the problem. Milt felt it had something to do with the landing

---

5. Whiteside interview by Lister for incident with the pilot.

gear, wondering if there wasn't enough damping in the oleo-type shock system. Ken Iliff suggested that maybe Milt was feeding the roll motions with the stick or rudder pedals. Absolutely not, replied Milt, adding that he had made sure during lift-off that he wasn't making roll or yaw control inputs.

We planned to get a little data for analyzing the problem by installing an instrumentation system in the M2-F1 after we returned from wind-tunnel testing at NASA Ames. Before that, however, using a ground-chase vehicle, we made some 16mm movies taken from the rear of the M2-F1, having painted references stripes on the rudders so we could determine their positions. The movies showed that the rudders were moving back and forth during lift-off. When Milt saw the movies, he concluded that slop and inertial weights in the rudder system—and not the pilot—were causing the rudders to move.

Larry Taylor suggested that we construct data from the movie frames. Using a stop-frame projector, we could determine right and left rudder positions and body roll angle on the M2-F1 by its position against the horizon in the background. We projected the filmed images of the M2-F1 onto a large sheet of paper we had hung on the wall. Using a protractor, we measured the roll angle and positions of both rudders in each movie frame. Using the frame rate of the projector, we then produced plots or time histories of the rudder movement and roll angle. Producing flight data in this way was hard, mundane work. Ken Iliff, Larry Taylor, and I took turns working with the data until we had in hand the results that Iliff and Taylor needed to analyze the problem.

In the hangar, we examined the rudder control system, finding it exceptionally stiff. No way could the rudders be moved without moving the pilot's stick. We examined the weight distribution of the rudder system, looking for how inertia could cause the rudders to move during vehicle roll. We still could not find the cause of the rudder motions.

Ken Iliff compared the phase relationships between rudder position and roll angle, giving Larry Taylor his findings. The control motions were typical of what a pilot would put in to combat roll oscillations. Finally, Larry Taylor and Ken Iliff put together a strong statement, saying they had no doubt that, knowingly or unknowingly, the pilot was working to combat the roll and that continuing to try to fly the M2-F1 with the control system driving the rudders from the pilot's stick would, during roll control, lead eventually to loss of control of the vehicle. They insisted that the current control system be abandoned and the other control system driving the elevons from the pilot's stick be hooked up for the next series of car-tow tests.

We couldn't share their conclusion and recommendation with Milt Thompson at the time, Milt being away on a trip. We had only one week left after Milt's return for car-tow tests before we were scheduled to go into full-scale wind-tunnel tests at NASA Ames. Given the strength of Taylor and Iliff's conviction about the control systems, I didn't want to waste time doing more car-tow tests with the original control system, so I asked Vic Horton to change the control system as Taylor and Iliff had recommended so that, when Milt returned, the M2-F1 would be ready for more car-tow tests.

After I made this decision, I noticed that the group had lost some of its harmony and camaraderie. Tension began to build between group members as they began to realize that a pilot's life could be at stake in this disagreement within the group over which roll-control system we used in the M2-F1. Milt Thompson was such a personable guy and worked so closely with us almost daily that emotions started emerging whenever critical decisions had to be made. I began to think that maybe it was better to have the research pilot more distant from the project people.

By backing Iliff and Taylor's recommendation, I had alienated Bertha Ryan and Harriet Smith to some extent. Ryan read me the riot act for not including Milt in the decision to change the control system, saying that, after all, it was his life at stake. I replied that Milt Thompson still had veto power as the pilot and that, if he insisted we do so, we would change back to the original control system. Ryan seemed satisfied by what I had said, but harmony on the project remained strained from that point. Even bigger conflicts would come later in the lifting-body program as the project grew.

As soon as Milt Thompson was back, I told him about the change made in the roll-control system. He was disappointed, wanting to do some more testing while using the previous control system, but he accepted the change, saying he still thought the problem was caused by the landing gear and that, when the new control hookup didn't solve the problem, we could go back to the original hookup.

With Milt Thompson onboard, we again hooked up the M2-F1 to the Pontiac and, with Whitey at the wheel of the Pontiac, off they charged across the lakebed. Cautiously, Thompson rotated the nose of the M2-F1 until there was very little weight left on the wheels. He continued to rotate the nose until the wheels were about three inches above the lakebed. The M2-F1 remained steady as a rock. We made another run, this time to an altitude of three feet. Thompson was gently maneuvering the M2-F1 right and left behind the Pontiac, but the lifting body showed no tendency to oscillate.

By now, Whitey had gone to Mickey Thompson's hot rod shop in Long Beach to replace the Pontiac's rear tires with drag slicks, a change that increased the car's towing speed to 110 miles per hour. Normally, drag racers use the wide, high-traction, threadless tires generally known as "slicks" because torque from the drive train to the lower gears is greatest at the start of the very short race known as a "drag," when tire slippage is most likely to occur. Our experience was exactly the opposite, with the height of drag found at the high-speed end of a tow. At about 90 miles per hour minus the slicks, the tires on the Pontiac would start slipping. Adding the drag slicks on the rear wheels of the Pontiac increased the towing speed enough to allow Milt Thompson to climb to twenty feet in the M2-F1, release the tow-line, and get about ten seconds of free flight before the flare landing.

Using the new control system, the M2-F1 handled well, both on and off tow in flight. Milt Thompson seemed to be happy with the control system. Neither Ryan nor Smith ever suggested later that we go back to the original control system. Not being an "I-told-you-so" sort of guy, I never again brought up the topic. And never again did we discuss control-rigging within the group, other than how to reduce stick forces with aft stick positions.

The Pontiac towed the M2-F1 for the first time on 1 March 1963, and before April was over, it had towed it a total of 48 times. While the Pontiac was prominently a part of the M2-F1 adventure, it was no secret that the car didn't exactly resemble the usual flight-line vehicle. According to Whitey, whenever someone from NASA Headquarters was visiting the Flight Research Center, Paul Bikle would slip away momentarily to phone him, telling him to hide the car. Whitey would pull the Pontiac behind a shed and throw a cover over it, the Pontiac "grounded" until the visitor left.[6]

What happened to the NASA muscle car once the M2-F1 program ended? Near the end of 1963, the Pontiac was shipped to NASA Langley Research Center in Virginia and used in tests at Wallops Island. There was some regret expressed at the NASA Flight Research Center when the Pontiac left, fairly much captured in a comment printed at the time in the ***X-Press***, the NASA newspaper at Edwards Air Force Base: "No longer can we drive along the lakebed and pass the airplanes in flight."[7]

---

6. Whiteside interview by Lister for the grounding of the Pontiac.
7. As quoted in Hallion, ***On the Frontier***, p. 150n.

CHAPTER 3

COMMITMENT TO RISK

For the 350-mile trip from Edwards Air Force Base to the NASA Ames Research Center at Moffett Naval Air Station in Sunnyvale on the southern end of the San Francisco Bay, we removed the "elephant-ear" elevons from the M2-F1 and loaded the vehicle on a flat-bed truck. The ten-foot width of the lifting body on the truck's bed caused it to be classified as a wide load, requiring two escort vehicles, one in front and one in back of the truck. The M2-F1 created some sensation along the route. The drivers had a lot of fun talking about it to the people who crowded around them on stops along the way, wanting to see it up close.

The NASA Ames Research Center is located in the heart of Silicon Valley, a few miles down the road from Stanford University. Moffett Naval Air Station had been the western operational base for Navy dirigibles in their heyday. The Navy dirigible ***Macon*** was a flying aircraft carrier, launching and recovering prop-driven fighter airplanes from its belly. The Navy was very proud of its dirigible fleet until two disasters happened: the ***Shenandoah*** crashed in an East-coast wind storm, and the ***Macon*** went down in the ocean off Monterey, California. Two hangars that housed these dirigibles still exist at Moffett. These hangars and the NASA Ames wind tunnel—its return section as tall as a ten-story building—are such prominent structures that they can be seen for miles by ground or air.

A bank of very large fans driven by electric motors generates the "wind" in the test section of the tunnel. Routed to the facility are power lines and a special substation. Operating the wind tunnel in those days required special coordination with the Edison Electric Company because of the need to have operators on standby to turn on extra generators when the tunnel was in use. To avoid conflict with peak daytime industrial electrical needs, wind-tunnel tests were often scheduled during night hours.

While it could take months or even years to get tests scheduled for this tunnel, Al Eggers had assigned a priority to the M2-F1 wind-tunnel tests. We had two weeks to conduct them. We had put together a test team consisting of both Vic Horton's hardware people and some of the analysis team. Horton participated in some of the data analysis, co-authoring with Dick Eldredge and Dick Klein the M2-F1 flight and wind-tunnel lift/drag results.[1]

---

1. Victor W. Horton , Richard C. Eldredge, and Richard E. Klein, ***Flight-Determined Low-Speed Lift and Drag Characteristics of the Lightweight M2-F1 Lifting Body*** (Washington, D.C.: NASA TN D3021, 1965).

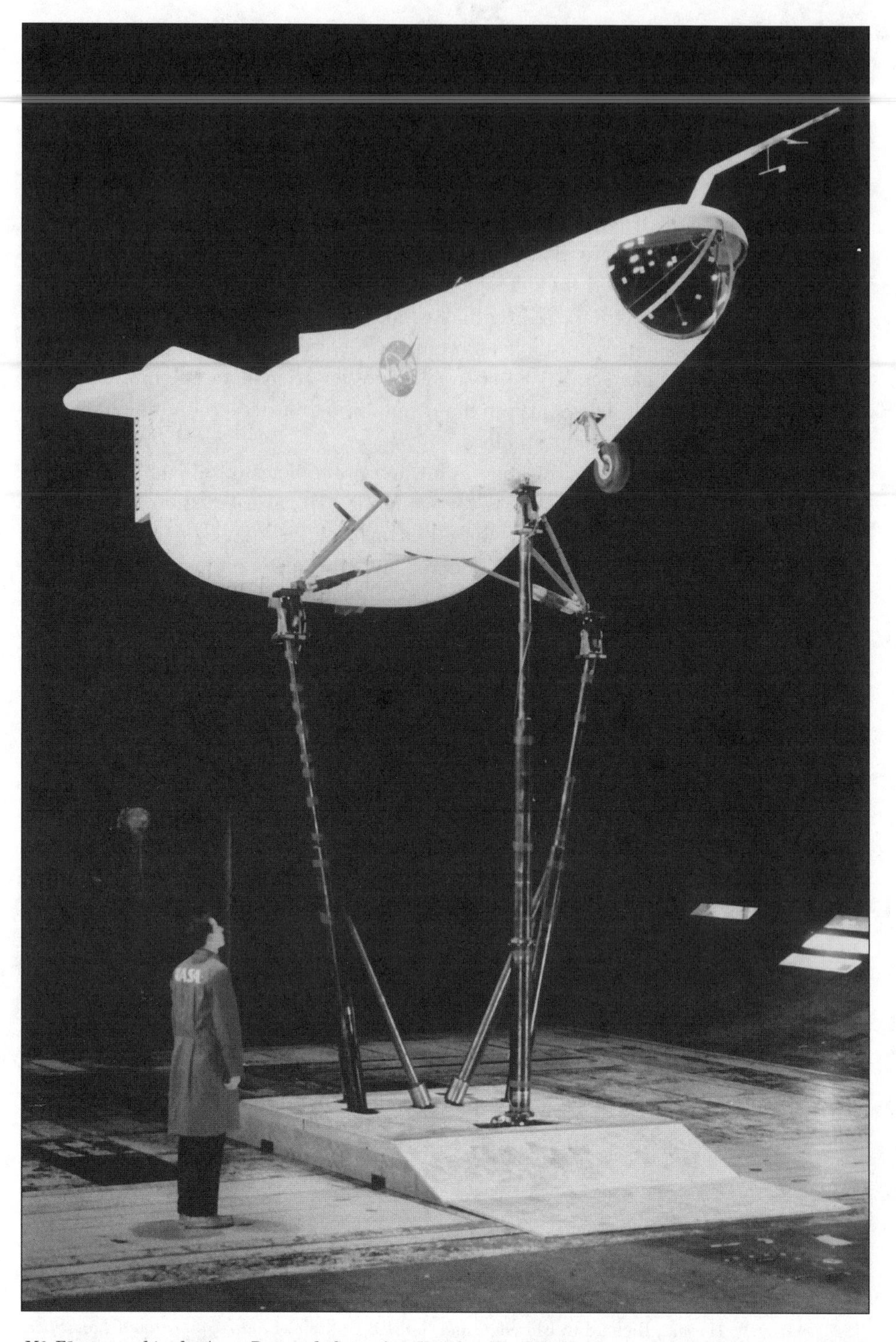

M2-F1 mounted in the Ames Research Center's 40X80-foot Wind Tunnel for testing. (NASA photo A-30506-15, also available as NASA photo EC97 44183-3)

While the NASA Ames crew operated the tunnel, our crew from the NASA Flight Research Center worked with the M2-F1—quite a different sort of adventure for a bunch of desert rats used to airplanes that fly in open sky over miles of sand and rock. I found a trailer park nearby where I could park my small travel trailer for the two weeks, having brought my wife and our two daughters along as well.

The inside of the wind tunnel was an awesome sight, especially at night. One night, as the M2-F1 team was preparing for a test, I took my family on a tour of the tunnel. We boarded an open-cage elevator on the ground floor, then rose through a darkness of steel beams and unlit open spaces to the floor of the dimly lit test section. The tunnel was a huge closed-circuit system in the shape of a race track, its entire length being about half a mile. Soot from engines stained the walls, making the interior of the tunnel dark and dingy, adding an eeriness to the atmosphere. My wife, Donna, said the tunnel would be a wonderful place to make an Alfred Hitchcock movie.

When we were ready to begin wind-tunnel testing, we had the M2-F1 hoisted high overhead by a crane, then lowered through a large hatch in the top of the test section. The vehicle sat 20 to 25 feet off the floor on top of three tapered poles resembling stilts that were mounted on a turntable balanced on the tunnel's floor, the M2-F1 attached near its landing gear to the poles.

What we did in testing the M2-F1 was unique, something that probably couldn't be done now due to NASA's emphasis upon safety. We didn't have remote controls on the M2-F1, even though most wind-tunnel models of vehicles have them. To move the testing along more rapidly, we talked the NASA Ames wind-tunnel crew into letting us take turns sitting in the cockpit, setting the pilot's controls at different settings by using plywood form boards. By keeping someone in the cockpit during the testing, the wind tunnel could be kept running, with necessary control changes made by the person in the cockpit. Otherwise, it would have taken a long time to get the tunnel's wind speed stabilized each time we started up again after shutting down the tunnel to make a change in control setting, angle of attack, or sideslip.

Ed Browne, Dick Eldredge, Milt Thompson, and I tried out the cockpit for size. I found it scary sitting up there over 20 feet off the ground inside a plywood barrel-like vehicle perched atop three spindly poles inside a dark cavern, shaking around as a windstorm screamed past at 135 miles per hour. I then decided that the best use of my time would be directing the tests in the safe confines of the wind tunnel control room. With the wind-tunnel operators and I peering at the wind tunnel pilots through thick windows in the tunnel walls, they felt like some kind of biological laboratory specimens under scrutiny.

We had an intercom system set up for communicating between the cockpit and the control room—a much better way to communicate, we felt, than holding up messages scribbled on paper to be read through the vehicle's canopy, especially when asking for help during sudden attacks of claustrophobia or because of a final call of one's bladder for relief. Whenever the wind tunnel pilots moved the controls or the wind velocity increased, they could feel the vehicle move as the poles supporting it flexed, an experience they all found disconcerting until they got used to it.

Milt Thompson, however, even wanted to conduct another wind tunnel test. As he said years later, "I tried to get them to attach a rope to it and let me actually try to fly it in the tunnel, but they wouldn't go along with that."[2] What Milt had wanted to do was sit in the cockpit of the M2-F1 on the floor of the tunnel, the tow-line tied upstream of the vehicle. However, the tunnel's crew was not very enthusiastic about Milt's suggestion, saying they could see the tow-line breaking and Milt and the M2-F1 ending up plastered against the turning vanes at the end of the tunnel. Even offering to attach slack safety lines during his "flight" did not keep the tunnel's crew from turning thumbs-down on Milt's request.

Before we started the formal data-gathering part of the tunnel tests, Milt found excessively large stick forces at aft stick positions while sitting in the cockpit and moving the controls around at different air speeds and body angles in the airstream. To minimize hinge moments, we had designed the outer elevons' pivot points to be slightly forward of the elevons' center of pressure. However, the trailing-edge body flap had been hinged at its leading edge, producing large hinge moments and stick forces. Using the wind-tunnel's fabrication shops, Vic Horton and his crew attached stand-off aluminum tabs on the body flap to help hold up the trailing edge, alleviating force on the stick. While the tabs didn't entirely eliminate the stick force, Milt considered it enough lessened to be tolerable.

There was another problem involving a phenomenon called a "Kármán vortex" that can also occur behind large trucks on the highway. A driver in a car at certain distances behind a truck in calm wind conditions sometimes can feel a "Kármán vortex" as the airstream whips back and forth. With the M2-F1, at certain airspeeds in the tunnel, a low frequency beat was being fed back to the vehicle's control stick. After taping tufts of yarn around the aft body and control surfaces of the M2-F1, we discovered that a large, oscillating Kármán vortex was coming off the body's base and beating against the body flap.[3]

The NASA Ames resident aerodynamicist, experienced in vortex flows, suggested that if we could change or disturb the base pressure slightly, we might be able to break up the single large vortex into a bunch of much smaller ones that would not beat so badly on the flap control surface. Once again, Vic Horton's crew went back into the shop, this time making two aluminum scoops and mounting them at the base on each side of the vehicle's body. The idea was to scoop air from the sides of the body into the cavity behind the base, thus increasing the base pressure and, we hoped, destroying the Kármán vortex. Milt climbed back into the cockpit, and we tested the M2-F1

---

2. Wilkinson, "Legacy of the Lifting Body," p. 54.

3. On the Kármán vortex, see Michael H. Gorn, ***The Universal Man: Theodore von Kármán's Life in Aeronautics*** (Washington, D.C.: Smithsonian Institution Press, 1992), pp. 23-24.

with the scoops. It worked. Having made two aerodynamic fixes to the vehicle, we were ready for the formal data-gathering portion of the wind-tunnel tests.

Dick Eldredge took the first shift, sitting in the cockpit and setting the controls with the plywood form boards. We were on a roll that day, cranking out data faster than we had before. Earlier, we had lost three days of our scheduled time in the wind tunnel while waiting anxiously as the tunnel's crew repaired its balance-data measuring system. After Eldredge had spent two hours in the cockpit, we asked him over the intercom if he would like someone else to take over. He declined. We asked him again every two hours until we had tested for eight hours straight with Eldredge in the cockpit, knowing he had only some water with him. Finally, after eight hours, Eldredge admitted that he was getting hungry and needed to go to the bathroom.

Data from the tunnel's measuring system came to us on tabulated sheets showing side, vertical, and aft force measurements as well as moments of roll, pitch, and yaw. The sheets also provided air speed, angle of attack, and sideslip. The M2-F1 "pilot" —whoever happened to be sitting in the cockpit during the test—also made notes regarding the control settings. We then correlated the data from the notes with that from the tunnel's measuring system.

The analytical team members hand-plotted on graph paper every single data point, using a room downstairs that had been set up for us. Hundreds of hours were involved in this work, each of us on the analytical team—Ken Iliff, Bertha Ryan, Harriet Smith, and myself—doing our share of the work. I think even Milt Thompson plotted a few points.

Whenever I saw the hardware crew had completed a task, I put its members to work plotting data as well. Once, when I did this, I didn't make myself too popular. They had been entertaining themselves with a game during a work lull, while the tunnel's crew was doing calibration checks on the measuring system. One by one, they were running across the tunnel floor, up the side of the curved floor, and putting a chalk mark as high on the wall as they could reach, the object of the game being to see who could make his mark the highest. After watching them for awhile, I had said, "If you guys aren't doing anything, come on down and help us plot data." Obviously, plotting data wasn't nearly as much fun as the game they had been playing, but they helped us anyway. A few years later, those marks were still on the tunnel's walls. Now, over thirty years later, I have often wondered if those marks are still there. If they are, they are probably covered up with additional layers of soot by now.

Some aspects of the good old days weren't so good, and one of them was having to spend those hundreds of hours hand-plotting data. Today, most wind tunnels have fully automated data systems with final plots rolling out of the machine soon after a tunnel test is finished. Today's engineer can analyze the data as it comes from the tunnel tests, modifying the test program in real time if an aerodynamic quirk shows up.

When our two-week stint at the NASA Ames wind tunnel ended, we packed up our data and trucked our little lifting-body vehicle back to its hangar at the NASA Flight Research Center. When we replaced the data in our simulator, based on the small-scale wind-tunnel tests, with the new data from the full-scale tests, we saw a difference. We knew that the only way to confirm the flight potential of the M2-F1 was to move on at once into actually flying it.

Gearing Up for Flight-Testing the M2-F1

Immediately after returning to the NASA Flight Research Center, we began planning how to move directly into air-towing the M2-F1 into flight. The tow-plane we decided to use was NASA's R4D utility aircraft, a Navy version of the Air Force's C-47, both being military versions of the legendary DC-3. Fondly dubbed the "Gooney Bird," the Douglas C-47 aircraft played a significant role during World War II as a glider tug during campaigns in Sicily, Normandy, and elsewhere. Now, the Gooney Bird was about to enter aviation history again as the tow-plane for the first lifting-body vehicle.

NASA's Gooney Bird was being used in several other ways, mostly as a transport aircraft. It had long been used at the Flight Research Center to shuttle people to and from Ames in support of joint activities. It was also being used in the on-going X-15 program to ferry people and equipment between Nevada lakebed emergency-landing sites and remote radar-tracking stations.

For a while we couldn't find a glider tow-hook for the Gooney Bird. Of World-War II vintage, this device was no longer in the military inventory. Finally, Vic Horton scrounged up one from a surplus yard in Los Angeles. We had no more than attached it to the tail of our Gooney Bird and run the release-line control up to the cockpit of the M2-F1, however, than we began to see dark clouds gathering over the lifting-body project as other people at the Flight Research Center began to believe that we were actually serious about flying the M2-F1.

First, Joe Vensel, local NASA Chief of Flight Operations, said that we couldn't fly the M2-F1 without installing an ejection seat. Eldredge and I told Vensel that we wished he had come up with this requirement when we were designing the vehicle. Fortunately, since the pilot sat at the center of gravity in the M2-F1, we found that we could add the ejection seat without unbalancing the lifting body. However, when we added the ejection seat and instrumentation, the M2-F1's weight rose to 1,250 pounds. To fly, the heavier vehicle required higher airspeeds than we had anticipated.

Because of this change, Dick Eldredge, Meryl DeGeer, and I went back over the structural load capacity of the M2-F1. We found that the most critical part of the structural design was the bending moment at the base of the vertical tails. The most severe flight condition, consequently, would be a high-speed dive in which the vehicle was forced into a high sideslip angle with the roll control (elevons) put in the wrong direction, adding to that bending moment. Using the simulator, we found that the only way a pilot could encounter that dangerous condition would be by attempting an aerobatic roll. A placard we added to the instrument panel in the cockpit clearly defined this limitation in four words: "No Aerobatic Roll Maneuvers."

At that time, the Weber Company was in the process of developing what we needed, a zero-zero ejection seat—that is, an ejection seat that operates even with the aircraft on the ground standing still (at zero altitude and zero velocity). The company was modifying a lightweight seat designed for the T-37 jet trainer to use a rocket rather than a ballistic charge for ejection. Joe Vensel came up with funds from his operations budget to pay Weber for this ejection seat to install in the M2-F1.

Very likely, the M2-F1 used one of the first zero-zero ejection seats ever made. Since Weber had not yet fully demonstrated the seat at the time, we arranged for a series of tests at the south lakebed where ejection seats were generally tested. Meryl DeGeer and Dick Klein worked with Weber in demonstrating and testing the seat.

Dick Klein constructed a plywood mockup of the M2-F1's top deck and canopy through which to fire a dummy sitting in the ejection seat. This dummy was fired up six times in the ejection seat. On each of the first five times, something went wrong and we had to make an adjustment.

Meryl DeGeer remembers Milt Thompson watching one of these tests. After the dummy and the seat smashed through the M2-F1 canopy mockup with rocket burning bright, the dummy separated from the seat at the top of the trajectory. The seat safely descended to the ground on a special parachute that Weber had added to save the seat for use in future tests. But the dummy, with its parachute still unopened, went sailing through the air head-first like Superman, its arms flapping.

As the dummy arched toward the ground, DeGeer glanced around at Milt Thompson. His face contorted, Milt was shouting at the dummy, "Flare! Flare! Damn you, flare!" The dummy ignored him and kept on flapping its arms as if trying to fly. The dummy crashed headlong into the bushes. Only then did its parachute flare open.

Everything worked well on the sixth test of the ejection seat, and the seat was installed in the M2-F1 without repeat testing to prove reliability. A year later, in 1964, an updated version of this seat was installed in the NASA Lunar Landing Research Vehicle (LLRV), the same seat that saved the lives of astronaut Neil Armstrong and pilot Joe Algranti when control systems failed in the Lunar Landing Training Vehicle at Johnson Space Center during training missions for landing on the moon.

Next, Thomas Toll, Chief of the Research Division, began to have serious doubts about flying the M2-F1. A respected but conservative researcher who had transferred to the NASA Flight Research Center from NASA Langley in Virginia, Toll had been one of the men responsible for the concept of the X-15. He felt that as long as we weren't flying the M2-F1 more than a few feet off the ground on car-tow, the data return was likely worth the effort, cost, and risk. Merely flying the M2-F1 on car-tow, he believed, would be a good learning tool for sharpening engineering skills in aerodynamics and stability and control, and it was also possible that the car-tow flights might even produce some useful data on lifting bodies.[4]

However, we were now thinking about flying the M2-F1 to high altitudes behind a tow-plane and that, he felt, was quite another matter. His serious misgivings seemed mostly to have to do with the fact that Milt Thompson had encountered a dangerous

4. For Toll's position, Hallion, ***On the Frontier***, p. 151.

lateral oscillation the first time he flew the M2-F1 on car-tow. Toll did not believe that any potential return in air-tow flight data was worth the risk to the pilot.

Toll had two other main reasons for opposing air-towing the M2-F1 into flight. First, he felt that the very low wing- or body-loading at which we were flying was not representative of a potential full-scale spacecraft, for an actual spacecraft the size of the M2-F1 would most likely weigh 10,000 to 15,000 pounds, ten times the weight of our M2-F1. Secondly, we weren't using any of the automatic control features, such as rate damping or automatic stabilization, that probably would be used in a spacecraft.

Paul Bikle tried to reason with Toll, assuring him that he felt it was worth the risk and that he would like to have Toll's endorsement. But Toll refused, going on record as refusing to endorse the planned M2-F1 air-tow operation. When Bikle went ahead and gave us the green light to proceed without the concurrence of NASA Headquarters or his own Chief of Research Engineering, he essentially was making a decision that could put his career with NASA on the line.

What Bikle did was an act of the kind of courage that I had never before seen in a manager. Essentially, he risked his career to support something that he believed in. There are basically two kinds of courage in the aerospace industry: the courage of test pilots who risk their lives, and the courage of managers who risk their careers to support decisions they believe are right, even when others disagree strongly. In his book ***The Right Stuff***, Tom Wolfe was correct to immortalize pilots as heroes.[5] On the other hand, program managers are responsible not only for the pilots who have "the right stuff" but also for the people involved in the program who have "the real stuff," as I have called it. When test pilots pay the ultimate price while risking their lives to test new aircraft, history remembers them as heroes who gave their all to aeronautical research. However, when program managers make a challenging decision simply because they believe it is the right thing to do, they risk being labeled failures or going down in history as bumbling idiots.

Today's program managers rarely encounter such risk, many of them using the bureaucratic process to build up walls that protect their careers. Today's manager can avoid risk by having decisions made by committees or by dividing programs into enough parts that it's not clear who is responsible for what. Another strategy that some program managers use to avoid risk is to be involved only with low-risk portions of a program, handing off high-risk portions to other managers who, if the program fails, can always defend themselves from blame by saying they were ordered to do the job. If the program succeeds, then the original program manager can step back into the picture and take credit for the successful venture by claiming it was his or her idea all along.

Paul Bikle would not have done well in today's managerial environment. He lacked the political imperative needed to work the system in his favor. He was so open and honest that everyone knew exactly what he was thinking—except when he was

5. Tom Wolfe, ***The Right Stuff*** (New York: Ferrar, Strauss, Giroux, 1979).

playing cards with the crew during lunch. That he was so open and "readable" was a trait that worked well for those working under him, for they knew where he stood. But it wasn't a trait that helped him in dealing with the hierarchy that developed gradually over him at NASA Headquarters.

Long after the lifting-body program was over, when Milt Thompson had retired as a research pilot and entered management at the Flight Research Center as its Chief Engineer, we talked about the episode with the Chief of Research Engineering back in 1963 and how people in different positions can view the same situation very differently, depending on their positions. As a research pilot, Milt Thompson had the reputation of being a wild and crazy guy who would take every calculated risk that his bosses would allow. But when he became a manager, he became very conservative, not allowing other pilots to take the same kinds of risks that he had taken as a pilot. In this sense, a manager is rather like the father who won't let his son ride motorcycles even though he had done so when he was a young man. As a manager, Milt Thompson said he could fully appreciate the position taken by the Chief of Research Engineering in vetoing the M2-F1 flight tests. He conjectured that if he had been in the Chief's position, he might also have questioned the rationale for the M2-F1 flights.

Gooney Bird Meets Flying Bathtub: First Air-Tow, 16 August 1963

After the ejection seat had been installed in the "Flying Bathtub," Milt Thompson made a few more tests on car-tow, adjusting to the heavier weight and checking out the flight instrumentation system. We did as thorough a flight readiness review as we could before moving into air-towing the lifting body, wanting to make sure there was not something we were overlooking.

One day while we were still getting ready to begin air-tows, Milt said to me privately that he had complete confidence in me to make the right decisions and that he was putting his life in my hands. That was the best and most sincere compliment I have ever received during my career.

By now, George Nichols and Glynn Smith, instrumentation technicians who had joined the lifting-body group of volunteers, had installed the instrumentation needed to radio data to the ground. Since the M2-F1 was an extremely simple glider with no onboard electronic systems, data from only 15 sensors would be sent to the ground. (By contrast, data from 400 to 500 sensors—later about 1100—was transmitted by radio during a typical X-15 mission.) In the M2-F1, the sensors would transmit air data, including airspeed, altitude, angle of attack, and angle of sideslip; vertical, side, and longitudinal accelerations; gyro data, including roll, pitch, and yaw rates; and control position data from the single elevator, two rudders, and two elevons.

Stability and control flight data would be transmitted by radio back to the antennae on the roof of the main NASA building that housed the control room, 10 miles from the lakebed take-off site. Here, Ken Iliff, Bertha Ryan, and Harriet Smith would watch plotting recorders equipped with ink pens generating traces of the data received from the measuring sensors aboard the M2-F1. The data would also be recorded on tape for analysis after the flight. Also in the control room during the flight would be the mission controller, research pilot Bill Dana, who, exactly two years later, would pilot the M2-F1 for the first time and then spend a total of ten years as a lifting-body pilot. But for the first air-tow flight of the M2-F1 on 16 August 1963, he would be on the ground, serving as that all-important link between the pilot in the cockpit and the engineers in the control room.

Developed for the X-15 program, the control room at the NASA Flight Research Center also contained two large plotting boards that drew the track of the aircraft on maps of the surrounding terrain, based on data received by the radar-tracking dish antenna atop the building. The control rooms later built at the NASA Johnson Space Center in Houston, Texas, for the first human space programs (Mercury, Gemini, and Apollo) were patterned after this control room at the NASA Flight Research Center.

For the first air-tow of the M2-F1 behind the Gooney Bird, we set up for take-off at the extreme south end of Runway 17, the longest lakebed runway on Rogers Dry Lake at Edwards AFB. We really didn't know how well the M2-F1 could make turns behind the Gooney Bird. An additional advantage of using the longest runway was that, if the tow-line broke or released, Milt could glide straight ahead, making a landing on the lakebed, Runway 17, using only half the length of the almost 15-mile-long lakebed. Piloting the Gooney Bird would be NASA X-15 pilot Jack McKay.

The plan was that when the Gooney Bird reached the north end of Rogers Dry Lake, McKay would make a large circle counterclock-wise over the lakebed while rising to an altitude of 12,000 feet. Once there, the M2-F1 would be released off the tow-line. Vic Horton would observe the M2-F1 flight from the small plexiglas dome atop the Gooney Bird, watching the M2-F1 in tow behind the Gooney Bird and keeping McKay advised on what was happening with it. I would be monitoring the flight from a radio van at the take-off site.

At seven o'clock on the morning of 16 August 1963, the winds were dead calm on the ground and only about five knots at 12,000 feet. A ladder was needed for boarding the M2-F1. Milt was assisted by the crew chief, Orion B. Billeter, since considerably care was needed to avoid stepping on the thin wooden skin of the vehicle's upper body deck. Once Milt was strapped into the ejection seat, his helmet radio was checked out. Then, the canopy was lowered and secured in place, and the ladder was pulled away. After the tow-line was hooked to the M2-F1, Billeter pulled on it while Milt checked the release hook. The procedure was repeated with the tow-line and release on the Gooney Bird.

NASA pilots Don Mallick (who would fly the M2-F1 four months later) and Jack McKay started and checked out the Gooney Bird's engines. Before take-off, McKay tried to avoid blasting Milt with too much dust from the lakebed. Ready to go, Milt gave a thumbs-up. After checking with the Edwards AFB control tower, the base's ambulance and fire truck, and McKay in the Gooney Bird, mission controller Bill Dana gave the go ahead for take-off.

M2-F1 in tow behind R4D "Gooney Bird," with the nose positioned high so the tow plane is visible through the nose window. (NASA photo E63 10962)

Gently easing the throttles forward on the Gooney Bird, McKay began to roll slowly down the lakebed. The Gooney Bird accelerated until its tail lifted off the ground. Very gently Milt lifted the M2-F1 off the ground exactly as he had done during the car-tows, slowly climbing on the end of the 1,000-foot tow-line until the M2-F1 was about 20 feet higher than the Gooney Bird and he could see the tow-plane through the nose window between his feet. He had to be fairly precise in maintaining position to keep the tow-plane in sight through the small nose window. The Gooney Bird gently lifted off the ground, Milt flying the M2-F1 in perfect formation behind and above the tow-plane.

After a few minutes of climbing, Milt radioed that the M2-F1 was very solid and that it was easy to hold high-tow position behind the Gooney Bird. Because we hadn't installed a pilot-adjustable pitch trim system, however, he had to hold back pressure on the stick. We had omitted doing that, just to keep it simple. The trim tabs we'd installed on the body flap during the wind-tunnel tests would trim out most of the stick forces in free flight but not on tow.

McKay held to a speed of 100 miles per hour as the Gooney Bird climbed to 12,000 feet. Over the radio, Milt said that he was beginning to relax and enjoy the flight. Nevertheless, he still had to give constant attention to keeping the Gooney Bird in sight through the nose window of the M2-F1. The Gooney Bird made three large circles over the northern lakebed during the twenty minutes taken to climb to 12,000

M2-F1 being air towed. Notice the side windows above the nose gear for increased visibility near touchdown. (NASA photo EC63 229)

feet. By this point, NASA pilot Fred Haise was flying alongside Milt in a T-37 jet trainer as a chase-observer.

The plan was for Milt to release the M2-F1 from the tow-line at this elevation while heading south over the northern portion of the lakebed. He was to make a 180-degree turn to the left, make a practice landing flare at about 9,000 feet altitude, and then push over and continue another 180-degrees to the left in order to line up on Runway 18, heading south. The average rate of descent was about 3,600 feet per minute, giving Milt about six minutes to learn to fly the M2-F1 before having to make the crucial one-shot landing maneuver.

Unlike the normal landing of an airplane, landing the M2-F1 was more like pulling out of a dive. A pushover maneuver had to be done at about 1,000 feet to build airspeed up to about 150 miles per hour, followed by a flare at about 200 feet altitude from a 20-degree dive. The flare maneuver would take about 10 seconds, leaving three to five seconds for the pilot to adjust to make the final touchdown. Milt had the option of hitting a switch to fire a rocket motor, giving him five to six more seconds to adjust sink rate before touchdown.

Watching from the ground, it seemed that the M2-F1 literally fell out of the sky. Since the vehicle had come level while Milt was making his practice landing at altitude, he radioed that he was going for the real one. Bill Dana, whose call sign was "NASA 1," confirmed that the practice landing had also looked good on the charts in the control room. Having made strip chart overlays earlier while Milt was practicing landings on the simulator, Ken Iliff, Bertha Ryan, and Harriett Smith had been able to do real-time comparisons while Milt was doing his practice landing maneuver.

However, if Milt hadn't been able to achieve level flight during the practice landing, our ground rules were that he was to eject, letting the M2-F1 crash. We considered the M2-F1 cheap enough to be expendable. Such a ground rule wouldn't sell in today's flight-testing of expensive airplanes, former NASA pilot and astronaut Fred Haise recently told me.

Our ground vehicles were parked well to the side of Runway 18, opposite Milt's planned landing point. It was scary watching him dive for the ground, and I held my breath. Milt leveled out, making a picture-perfect landing at the planned touchdown spot without using the rocket. I finally remembered to breathe as he rolled straight ahead and turned off the runway, coasting to a stop. All of us, including Paul Bikle, surrounded the M2-F1 while Orion Billeter helped Milt out of the lifting body. We were one bunch of happy people as we stood there, shaking Milt's hand. Later, the debriefing room was wall-to-wall smiles as Milt described a flight that went exactly as planned.

We had a party that night at my house, but it bore no resemblance to the typical wild X-15 parties of heavy drinking that Milt Thompson described in his book, ***At the Edge of Space***.[6] Since the X-15 program involved most of the personnel of the Flight Research Center in some way, X-15 parties were always held at Juanita's, then the biggest bar in Rosamond, just outside the western boundary of Edwards AFB. Almost exclusively stag, the X-15 parties were mostly attended by NASA's ex-military pilots, aircraft crews, and flight planners. As Milt relates, most of the X-15 parties continued at Juanita's for four or five hours, then moved to one or more of the bars in Lancaster.

Unlike the X-15 program, the lifting-body program had research engineers steering its path from the very beginning. After success with the M2-F1, additional lifting-body vehicles would continue to be designed and built throughout the twelve years of the lifting-body program, involving the cooperation and teamwork of research and design engineers at three NASA centers (including Ames and Langley) as well as the research engineers at contractors Northrop and Martin.

The lifting-body program was also the first program at the NASA Flight Research Center significantly influenced by women engineers. Bertha Ryan and Harriet Smith not only played major roles in the development of the M2-F1 but continued to do so with other lifting-body vehicles, by which time other women at the NASA Flight Research Center were also involved in the program. Afterwards, Harriet Smith moved on to project management at the NASA Flight Research Center, while Bertha Ryan opted to remain in research engineering, later designing missiles for the Navy at the China Lake Naval Weapons Center, about 50 miles north of Edwards AFB. Since the days of the lifting-body program that ended in 1975, women have increasingly entered the world of aerospace technology, so that now it is common to see women in engineering, as part of flight crews, and as pilots and astronauts.

---

6. Thompson, ***At the Edge of Space***, pp 71-73.

The successful flight of the M2-F1 was a special triumph for us, a little team of "amateurs" pulling off a big one. Despite having Paul Bikle's full backing, many of the "professionals" on the X-15 program had continued to consider the M2-F1 a high-risk project due to our lack of experience. Of course, it's a little hard to have much experience when doing something that has never been done before. We matured at once after that first flight, rapidly moving up in credibility and status.

A few weeks after the successful flight of the M2-F1, on 3 September 1963, aviation news reporters first viewed the craft at the NASA Flight Research Center. The M2-F1 quickly became a hot item in aviation periodicals.

While a few people at NASA Headquarters were aware of the lifting-body project until about a week after the historic first flight of the M2-F1 in mid-August, they did pay much attention to it, mainly because we hadn't requested any money for the program. However, the NASA administrator in Washington, D.C., James E. Webb, remained unaware of the successful first flight of a lifting body until, while testifying before a congressional committee, he was asked about it by a congressman who, having read about it in the press, wanted to know if NASA was starting a new multi-billion-dollar space program that Congress neither knew about nor had approved. Bikle's phone began ringing immediately after this incident, which obviously had been embarrassing for the administrator. When Webb found out that we had spent only about $30,000 on the program and that there was no billion-dollar plan in the making, things cooled down and we were allowed to continue with our M2-F1 flight tests.

December 1963: Peterson, Yeager, and Mallick Fly the M2-F1

After Milt's first flight, the M2-F1 became very operational. As a simple glider, it had no systems to maintain, except the research instrumentation system. A part-time crew chief could easily keep the M2-F1 on flight status. The Gooney Bird was available most of the time to us as a tow-plane because it was being flown almost every day on support missions for other programs and had a full-time crew chief.

Milt Thompson flew the M2-F1 on its first seventeen flights in 1963—five in August, two in September, six in October, three in November, and one in December. These flights were made specifically to define the craft's aerodynamics and stability and control characteristics. Flight research is most valuable when the data is used, as it was in these first flights, in comparison with wind-tunnel test results in correcting or completing aspects of design and prediction based on those results.

After these flights, Paul Bikle and Milt Thompson decided it was time to start checking out other pilots in the M2-F1, beginning with Bruce Peterson and Colonel Chuck Yeager. A NASA test pilot and a former Marine Corps pilot, Bruce Peterson had served along with Milt Thompson in 1962 as one of two project pilots on the paraglider research vehicle, or Paresev, program that was designed to evaluate the use of an inflatable flexible wing in the space program as a way by which astronauts could leave a spacecraft and return to Earth in a vehicle capable of making an airplane-like landing. The similarity with the M2-F1 is that both vehicles were gliders towed into

M2-F1 pilots (Chuck Yeager in cockpit, Bruce Peterson to his left, and Don Mallick) being checked out by Milt Thompson (on stool). (NASA photo E63 10628)

flight by winged aircraft, and both programs were excellent examples of Paul Bikle's low-cost and do-it-quick approach. Paul Bikle wanted his old friend, Chuck Yeager, then head of the USAF Test Pilots School at Edwards AFB, to fly the M2-F1 and give his assessment of the vehicle before other Air Force pilots were allowed to fly it.[7]

During the last week in November, Peterson and Yeager were initially checked out on the M2-F1 during extensive car-tows up to an altitude of 20 feet. Thompson scheduled both for flights in air-tow by the Gooney Bird on 3 December, using a five-mile-long lakebed runway so that there would be nothing critical about where touchdown occurred on the runway so long as a good flare was made to keep from breaking the M2-F1 in hard landing. With Bill Dana and Don Mallick piloting the tow-plane, Peterson and the M2-F1 were towed aloft to 12,000 feet in the first flight of the day. Peterson released the tow-line, making a very good landing on the lakebed. However, the M2-F1 had landed some distance from the van containing Milt Thompson and Chuck Yeager, which was sitting beside the runway.

---

7. For further information on the Paresev program, see Hallion, ***On the Frontier***, pp. 137-140; Lane E. Wallace, ***Flights of Discovery: 50 Years at the NASA Dryden Flight Research Center*** (Washington, D.C.: NASA SP-4309, 1996), pp. 131-33.

Next, it was Yeager's turn to have his first lifting-body flight. Naturally competitive, Yeager suggested going for a spot landing on the runway just opposite the van parked beside the runway. Dana and Mallick towed Yeager aloft, as they had Peterson. Yeager opened up the flight envelope on the M2-F1, flying both faster and slower in his practice landing maneuver at altitude than had Milt. Then, he dove the M2-F1 at the lakebed in a steeper angle than Milt had used, leveled out, and made a greased-on landing in front of the van. Climbing out of the M2-F1, Yeager exclaimed, "She handles great!"[8]

It was a beautiful, but cold, December morning. The winds were still calm, and the Gooney Bird had been climbing very well in the cold weather. Milt suggested that Peterson and Yeager each get two more flights in for the day. Responding to Yeager's challenge, Peterson set up in his second flight to touch down just in front of the van.

What Peterson and the rest of us didn't realize was that we engineers had made a little mistake. Since Milt had started flying the M2-F1 in August and the weather had been quite warm whenever he flew the vehicle that fall, we had serviced the shock struts in the main landing gear with a standard viscosity oil. That was fine for Milt's earlier flights. However, on this early December morning, after two flights to altitude in temperatures below freezing, the oil had hardened to the consistency of molasses.

When Peterson landed the M2-F1, the landing gear was rigid, the struts immobized by the thickened oil. At touchdown, the main wheels separated from the vehicle and bounced across the lakebed, as shown in the film of the landing made by the forward-looking camera mounted behind the pilot's head. The four bolts connecting the wooden shell to the inner steel tubing also tore out, dropping the wooden shell about six inches until it settled around Peterson in the cockpit.

Not injured, Peterson was the brunt of jokes about this landing for years afterwards. Structural repairs were easily made to the M2-F1. The original Cessna 150 landing gear was replaced with the more rugged gear of a Cessna 180. Different struts were added with multiviscosity oil. Before continuing flights nearly two months later in late January 1964, we expanded the research data system to measure more parameters for extraction of aerodynamic derivatives.

The first flights of the new year were made on the morning of 29 January, with Bruce Peterson, Milt Thompson, and Chuck Yeager each making two flights. Yeager said he was having a ball flying the vehicle. The next morning, NASA pilot Don Mallick checked out in his first two and only lifting-body flights after Yeager made his fourth and fifth (last) flights in the M2-F1.[9]

During the briefing session before the day's flights, I had denied Yeager's request to be allowed to roll the M2-F1. He believed that he could make a perfect barrel roll in the little lifting body. I explained that Dick Eldredge and I had designed the M2-F1 to weigh 800 pounds and fly at a slower speed, not knowing the vehicle would

---

8. Quoted in Hallion, ***On the Frontier***, p. 152.

9. The five flights include only the air tows, not those in the M2-F1 towed by the Pontiac.

have to grow in weight to 1,250 pounds by adding an ejection seat, heavier instrumentation, and a landing rocket. I also explained that we weren't that confident in analyzing loads in a roll maneuver, for not only were there bending moments from side loads in sideslips, but loads also were transmitted to the vertical tails from the asymmetrical "elephant ears" attached to them. In short, we couldn't be sure the tail would remain intact during a roll, given the vehicle's heavier weight.

Yeager didn't try to roll the M2-F1 on his last flight that morning. As experience later showed, however, Yeager likely could have barrel-rolled the vehicle successfully that morning, for over a year later the M2-F1 was rolled unintentionally in two flights and the tail remained intact. Although Yeager never flew a lifting body after his fifth flight in the M2-F1, he remained very enthusiastic about the concept, exerting a good deal of influence in encouraging the Air Force to develop the rocket-powered lifting bodies, the X-24A and X-24B, and the jet-powered X-24J.

Serious Research Flying, 1964-1965

After January 1964, we settled down into a year of serious research flying. Milt Thompson and Bruce Peterson often alternated as pilot, the M2-F1 flown a little over twice a month on average, as quickly as the research analysis team could digest data from one flight and plan the next. We made a total of 29 flights, 11 of them by Peterson.

Working together, Ken Iliff, Bertha Ryan, and Harriet Smith had put together a planned program for extracting data from three basic types of flight maneuvers—the steady state, quasi-steady state, and dynamic. In a typical steady-state maneuver, for example, the M2-F1 would be flown straight ahead and stabilized at different airspeeds in the glide, resulting in data for Jon Pyle and Ed Saltzman on lift, drag, and elevator trim. In a typical quasi-steady-state maneuver, the pilot would put the M2-F1 into a gliding wind-up turn and gradually tighten the turn, increasing the "G" load (gravitational pull) by increasing back stick pressure, allowing lift, drag, and trim data to be measured at higher airspeeds and with structural deflections, if any.

In a typical dynamic maneuver, the pilot would stabilize the M2-F1 in a steady glide and then pulse one control at a time, with the pulse usually in a doublet. For example, if the goal was to get data on aileron characteristics, the stick would be moved to the right and held, moved to the left and held, then returned to neutral and held fixed by the pilot. Then, the vehicle would be allowed to oscillate with controls frozen by the pilot. This maneuver would be repeated for several airspeeds or angles of attack, researchers extracting aileron characteristics from the doublet portion of the maneuver and airframe characteristics from the final portion of the maneuver involving oscillation with controls frozen. This maneuver was also done for defining yaw control by rudder and pitch control by stick fore and aft.

For the aerodynamic characteristics of the M2-F1 to be defined completely, Thompson and Peterson had to perform almost 100 maneuvers. With only about six minutes available as the M2-F1 glided down from 12,000 feet, the pilots used flight cards to squeeze in as many maneuvers as possible before having to set up for landing. Each flight averaged four maneuvers during those six minutes.

Aerodynamicists define the characteristics of a given airplane shape by the use of aerodynamic derivatives coming from three types of air forces: those caused by wind flow direction, angle of attack, and angle of sideslip; control deflections; and rotary motions. While there was plenty of wind-tunnel data on the M2-F1 to compare with flight data on the first two types of air forces, there was no wind-tunnel data for the third, those air forces caused by rotary motions of the vehicle. The first two types could be evaluated easily in the wind tunnel with the model held stationery on strings or pedestals. However, the third type can be evaluated only by using elaborate mechanisms to rotate the model rapidly in all axes (roll, yaw, and pitch). No attempt was made at NASA Ames to obtain this type of dynamic or damping data during the wind-tunnel testing of the M2-F1, not only because of the huge expense involved in developing the mechanisms, but especially because of the lack of confidence in this type of wind-tunnel data, the elaborate mechanisms interfering with the airflow around the model.

We "guesstimated" the rotary data that we put into the simulator along with the other data resulting from wind-tunnel measurements. Often, these "guesstimates" turned out to be off by a factor of three or four since, at the time, we didn't have good techniques for estimating aerodynamic rotary damping derivatives. Ken Iliff and Larry Taylor put their heads together, trying to come up with a solution.

They decided to convert Taylor's garage at his home in Lancaster into a wind tunnel for measuring rotary derivatives, using the original small-scale model of the M2-F1 that I had built. Taylor built a long box with a five-horsepower electric fan in one end, plus straightening vanes and a special test section in the middle. They sealed the garage door so the entire garage could be used to return the air to be recirculated through the box inlet, thus making it a more efficient closed-loop tunnel. Taylor also designed and rigged a balance system composed of strings, pulleys, and a very sensitive string tension measuring device so the M2-F1 model could be rolled, yawed, or pitched at different rates. Of course, lightweight household objects hanging in the garage had to be anchored to keep them from blowing around in the garage. These at-home wind-tunnel tests provided the data for the simulator estimates and for comparison with actual flight data.

Iliff and Taylor also applied a trial-and-error technique, originated by Dick Day on the X-2 project, that used the analog flight simulator for extracting derivatives in flight. They changed settings on the simulation one at a time until they got time histories of dynamic maneuvers from the simulator to match up with those recorded from flight. Although this was a long and tedious process with limited accuracy, it was the only way we knew at the time for doing this task with analog systems. Harriet Smith was primarily responsible for extracting derivatives from the M2-F1 flight data, using this technique with the analog simulator. In 1965, Smith published a report entitled "Evaluation of the Lateral-Directional Stability and Control Characteristics of the

Lightweight M2-F1 Lifting Body at Low Speeds," showing the flight results with wind-tunnel comparisons.[10]

Iliff and Taylor also had new tools coming into use by which to sharpen their trade, for the digital computer revolution was in full swing by this time. Within a few years, they developed a new computer technique called "the maximum likelihood estimator," by which dynamic-maneuver flight data could be input into a digital computer to produce aerodynamic derivatives—a technique producing very accurate results so long as the flight data used is high in quality and accuracy. In fact, "the maximum likelihood estimator" that Iliff and Taylor originated during the lifting-body era at the NASA Flight Research Center is currently being used by flight-test organizations in the United States and in various countries around the world.[11]

Aerobatics in the Flying Bathtub

Over the next two years, 1965 and 1966, the M2-F1 was used primarily to check out and familiarize more pilots with the lifting body, including NASA pilots Bill Dana and Fred Haise and Air Force pilots Joe Engle, Jerry Gentry, and Don Sorlie. By this time, flying the M2-F1 was also a kind of preparatory task undertaken by pilots who hoped later to fly the M2-F1's heavyweight successor, the M2-F2. The M2-F1 made 28 air-tow flights during 1965 and 1966, and by the time the first lifting body was retired from flight in August 1966, it had been flown by ten pilots about 400 times by car-tow and 77 times by air-tow. Fred Haise and Joe Engle flew the M2-F1 only on car-tows to 25 and 30 feet in altitude on 22 April 1966, their experience with the lifting body cut short due to their being selected as astronauts for NASA space missions.

Milt Thompson and Vic Horton developed a formal lifting-body pilot checkout procedure that required each pilot to make 24 car-tows before his first air-tow flight. The first three car-tows involved nose-gear steering with tow-line releases at up to 45

10. Harriett J. Smith, ***Evaluation of the Lateral-Directional Stability and Control Characteristics of the Lightweight M2-F1 Lifting Body at Low Speeds*** (Washington, D.C.: NASA Technical Note D-3022, 1965).

11. Another name for "maximum likelihood estimator" is parameter estimation, which can also be described as a series of mathematical procedures developed by Dryden researchers to extract previously unobtainable aerodynamic values from actual aircraft responses in flight. This contribution allowed flight researchers for the first time to compare certain flight results with predictions. A discussion of this technique appears in Lawrence W. Taylor and Kenneth W. Iliff, "A Modified Newton-Raphson Method for Determining Stability Derivatives from Flight Data," paper presented at the Second International Conference on Computing Methods in Optimization Problems, San Remo, Italy, Sept. 9–13, 1968. On this matter, see also Kenneth W. Iliff, "Parameter Estimation for Flight Vehicles," ***Journal of Guidance, Control, and Dynamics*** , vol. 12 (Sept.–Oct. 1989): 609–22.

miles per hour. The next six car-tows involved nose-wheel rotations up to 60 miles per hour. The final fifteen car-tows involved doing lift-offs at up to 95 miles per hour to familiarize the pilot with roll control with elevons and yaw control with rudders. Although the number of car-tows required seemed excessive to the pilots, Milt Thompson felt the requirement was necessary to minimize the risk of injury to a pilot or damage to the vehicle during car- and air-tows.

Before moving on to air-tows, the pilots were also familiarized with the flare portion of lifting-body flight by means of a three-degrees-of-freedom simulator and a shadowgraph presentation. This pre-flight procedure included familiarizing the pilots with the capabilities of the landing-assist rocket.

On 16 July 1965, it was Captain Jerry Gentry's turn to get checked out in the M2-F1. An Air Force test and fighter pilot who later made the first flight of the Air Force's X-24A and then flew missions in Vietnam, Gentry found flying the M2-F1 on air-tow to be challenging.

The lifting body was hooked by tow-line onto the Gooney Bird, and the takeoff began. Gentry lifted the M2-F1 into formation above and behind the Gooney Bird on the end of the 1,000-foot tow-line. Then, the Gooney Bird, piloted by Fred Haise, lifted off. At about 200-foot altitude, while Gentry was climbing, something began to go wrong. Gentry began making small roll inputs to correct the right and left positions of the lifting body relative to the Gooney Bird, his corrections growing larger and larger. All at once, we had another pilot-induced oscillation in the making.

As the amplitude of the oscillation increased, so did the urgency of radio contacts with Gentry:

"Level your wing..."

"Level your wings!"

"Release..."

"Release!"

"Eject!"

"Eject!"

As the Gooney Bird slowly climbed to 300 feet above the lakebed, Vic Horton was watching the M2-F1 through the tow-plane's observation dome, the rocking motion of the M2-F1 growing larger and larger. He watched in horror as the M2-F1 rolled belly-up and disappeared from sight below the tail of the Gooney Bird. Both Gentry and the safety monitor aboard the Gooney Bird released the tow-line, realizing the situation was completely out of control. Vic Horton was convinced that he'd next see pieces of the M2-F1 scattered across the lakebed, which, had it happened, could have been the end of the lifting-body program.

When it was released from the tow-line, the M2-F1 was inverted with its nose high and traveling at approximately 100 knots airspeed, or about 115 miles per hour. The probability of recovery from that condition was virtually zero. During a normal landing, with the vehicle straight and level, the flare would be initiated at that 300-foot altitude at a stabilized speed of 120 knots, or 138 miles per hour. Theoretically, at least, it was impossible to get the nose down in time to pick up the speed needed to

M2-F1 dummy ejection seat test setup at South Edwards. (Air Force photo JN-043-1, available as NASA photo EC97 44183-4)

accomplish flare. Fortunately, Gentry ignored theory and, after release from the tow-line, completed the barrel roll, touching down on the lakebed at the bottom of the roll...all in nine seconds. It was a hard roll that broke the landing gear but it produced no other damage or injuries, except to Gentry's pride.

Gentry was so upset that he insisted on trying another flight immediately. Other members of the operation, including the instructor pilot, were in such a state of shock at the time that they agreed to try again, even though the M2-F1 was obviously listing heavily to one side due to its broken landing gear. Luckily, cooler heads had observed the entire incident from the office of the Director of Flight Operations. A stern call came over the radio to knock it off and get back in here.

During the next thirteen months, while Gentry practice more car-tows, the repaired M2-F1 was flown nine times by Milt Thompson. On 16 August 1966, Gentry got his second chance at a checkout flight on the M2-F1. None of us expected history to repeat itself, but it did. We watched in shock as the same sequence of events rapidly developed, a low-amplitude lateral oscillation beginning immediately after liftoff, rapidly building to greater than plus or minus 180 degrees. Once again the tow-line was released with the M2-F1 upside down at 300 feet above the lakebed. Gentry must have found something familiar about the episode, for this time he released the tow, completed the barrel roll, came wings level, ignited the landing rocket, and made a perfect landing.

The second time around there was no damage to the vehicle, but Bikle apparently had had enough. "That's it!" he said. Bikle saw towing as a special problem with the M2-F1, believing that in future we should look into launching lifting bodies from bombers. Bikle stuck by his decision, grounding the M2-F1 permanently. With that, the first lifting-body vehicle was retired from flight.

Gentry later was able to prove to Bikle, Thompson, and Lieutenant Colonel Don Sorlie, the official boss of the Air Force's lifting-body pilots, that his problems in the M2-F1 were caused simply by a lack of visibility. Being much shorter than the other pilots affected his eye position in the cockpit of the M2-F1 considerably, so much so that after the lifting body and the Gooney Bird left the ground on tow, he could see neither the Gooney Bird nor the horizon through the nose window, making it humanly impossible to control the vehicle's attitude.

Recalling the event years later, Gentry said with a laugh, "Oh, hell, I was upside down twice on tow. As soon as I could figure out which way the roll was going, I put stick in with the roll and went on around. When I got momentarily to wings-level, I punched off. Barely had time to release the tow, flare, and whump. The second time it happened, I said, 'Well, I've been here before.' I'd gotten good enough at it that I even glided for a few seconds."[12]

Whether Gentry would be allowed to continue flying lifting bodies in future phases of the program rested entirely on the ruling of Bikle and Thompson after conferring with Sorlie. They decided to allow Gentry to continue with the rocket-powered lifting bodies. In 1992, in his acceptance speech at a Test Pilots' Walk of Honor awards ceremony in Lancaster, Gentry expressed appreciation for Bikle and Thompson's decision. While he was flying the lifting bodies, Gentry was the project pilot on the F-4E and later did flight tests on other aircraft including the F-4C/D, F-104, F-111, and F-5.

Significance of the M2-F1 Program

The M2-F1 program proved to be the key unlocking the door to further lifting-body programs, including the current Shuttle spacecraft and several other vehicles currently in-progress, such as the X-33. Flight tests of the M2-F1 supplied the boost in technical and political confidence needed to develop low lift-to-drag-ratio, unpowered, horizontal-landing spacecraft.

Technical reports written by the engineers who were part of the M2-F1 program also were important, establishing the lifting-body as a concept. For example, "Flight-Determined Low-Speed Lift and Drag Characteristics of the Lightweight M2-F1 Lifting Body" by Victor W. Horton, Richard C. Eldredge, and Richard E. Klein compared wind-tunnel and flight data to establish the fact that a lifting body with a maximum lift-to-drag ratio of 2.8 measured in-flight could be landed successfully and

12. Wilkinson, "Legacy of the Lifting Body," pp. 54–55.

repeatedly by an unassisted pilot.[13] Furthermore, the findings of the M2-F1 stability-and-control engineering team—Ken Iliff, Bertha Ryan, Harriet Smith, and Larry Taylor—demonstrated that a radically-shaped flying machine such as the M2-F1 did not need automatic control augmentation to have acceptable and even good handling qualities, a conclusion confirmed in Harriet J. Smith's "Evaluation of the Lateral-Directional Stability and Control Characteristics of the Lightweight M2-F1 Lifting Body at Low Speeds."[14]

In the 1960s, the lifting-body concept was so tentative in the minds of space planners that the M2-F1 program seemed destined to have pronounced effect on the direction taken afterwards in space vehicles, the potential for development of horizontal-landing spacecraft fairly much dependent upon our success. Any one of three different effects could have followed from the three major outcomes possible for the M2-F1 program:

First, the M2-F1 program could have halted after the car-tows at very low altitudes, as would have happened had Paul Bikle agreed with the then Chief of the Research Division at the NASA Flight Research Center. Had this happened, the expressed lack of confidence in the flight concept could have prevented the later acceptance of any proposed follow-on lifting-body programs, which, in turn, could have slowed or prevented the later development of a horizontal-landing spacecraft such as the current Shuttle.

Second, if we had had a serious accident with the M2-F1 in which a pilot was injured severely or killed, it isn't likely that any additional lifting-body flight-test programs would have taken place, making even less likely the later development of today's Shuttle and other horizontal-landing spacecraft.

Third, the M2-F1 program could be an adventure in success and open the door for future lifting-body programs. Fortunately, this is exactly what happened. And the door remains open for generations yet to come of the progeny of the original lifting body, the lightweight M2-F1.

---

13. Victor W. Horton , Richard C. Eldredge, and Richard E. Klein, ***Flight-Determined Low-Speed Lift and Drag Characteristics of the Lightweight M2-F1 Lifting Body*** (Washington, D.C.: NASA TN D3021, 1965).

14. Harriett J. Smith, ***Evaluation of the Lateral-Directional Stability and Control Characteristics of the Lightweight M2-F1 Lifting Body at Low Speeds*** (Washington, D.C.: NASA Technical Note D-3022, 1965).

CHAPTER 4

ON TO THE HEAVYWEIGHTS

When Paul Bikle grounded the M2-F1 permanently in mid-August 1966, a ground swell of interest in lifting-body re-entry vehicles had been growing for over two years within NASA. The initial flights of the M2-F1 had shown that the lifting-body shape could fly. As early as two weeks after the first car-tows of the M2-F1 in April 1963, Bikle had shared his confidence in lifting bodies with NASA Headquarters, writing Director of Space Vehicles Milton Ames that the more the Flight Research Center got into the lifting body concept, the better the concept looked.

Bikle also mentioned that he was noticing "a rising level of interest" in the lifting-body concept at the Ames and Langley centers. By 1964, NASA Headquarters and these two Centers had considerably increased their participation in the lifting-body concept through the Office of Advanced Research and Technology (OART) under the direction of NASA Associate Administrator Raymond Bisplinghoff.

By this time there were also many lifting-body advocates within the aerospace industry and the Air Force. The successful flights of the M2-F1 had accelerated the aerospace community's interest in the possibility of applying the concept of lifting re-entry to the next generation of spacecraft. After our M2-F1 success, the lifting body quickly rose toward the top of the Air Force's priorities in re-entry designs. Although there were still many in the Air Force holding out for variable geometry wings and jet engines to assist in landing recovery, wingless and unpowered vehicles had become more prominent in both NASA and Air Force studies.[1]

Change in Plans: On to Rocket Flight

I had originally planned to fly three lightweight lifting-body shapes. Once the M2-F1 had been built, I was ready to move on to the other two shapes, the M1-L and the lenticular. By this time, however, interest in building the other two shapes into vehicles had waned, replaced by the urge to fly a rocket-powered lifting body at transonic speeds.

After the M2-F1 was built in 1962, I went to NASA Ames and NASA Langley to confer with other engineers about developing lifting re-entry configurations. At NASA Langley, I discovered that the leading lifting-body advocates among the engineers were rapidly making progress. Eugene Love was leading the lifting-body interest at NASA Langley, with Jack Paulson, Robert Rainey, and Bernard Spencer conducting

1. Entire three paragraphs above, including quotation, based upon Hallion, *On the Frontier*, pp. 151-53.

studies and wind-tunnel tests on candidate designs. Although they were still considering deployable wings and jet engines, a powered and wingless lifting-body configuration—the HL-10 (for Horizontal Lander)—emerged as a strong contender after our success in flying the M2-F1.

NASA Headquarters assigned Fred DeMerritte as program manager for coordinating lifting-body activities at various sites, including the Flight Research Center, Ames, and Langley. We felt fortunate to have DeMerritte as program manager, for he was a good team worker who listened to us. His skill in cutting through red tape helped us move the lifting-body program along. We set up a planning team composed of three members, one for each of the three NASA sites. While I represented the Flight Research Center, George Kenyon and Bob Rainey represented Ames and Langley, respectively. Since Johnson Space Center in Houston, Texas, had become the leading Center for manned space exploration, the planning team met at Johnson with its representatives.

After Kenyon, Rainey, and I presented the views of our colleagues, we quickly narrowed in on two important objectives. First, future flight tests on lifting bodies should be at wing loadings or weights five to ten times more than those of the M2-F1. Secondly, flight-test vehicles should be capable of the higher speeds in the transonic and lower supersonic speed ranges where large changes in lifting-body aerodynamics occur.

After my return to the NASA Flight Research Center, I put together a plan for a heavyweight M2-F1 with the same dimensions as the original vehicle, proposing to launch the heavyweight version from the Center's B-52 in a way similar to how the X-15 was launched. The X-15 program was no longer using the LR-11 rocket engines, and we could use them now in our lifting-body program. The LR-11 engine consists of four separate barrels or chambers, each barrel developing some 2,000 pounds of thrust for a total thrust of about 8,000 pounds. The pilot had four increments of throttling capability since each barrel could be operated separately. Two LR-11 engines in the X-15 had achieved 16,000 pounds of thrust, the engines burning a combination of water and alcohol, with liquid oxygen employed as the oxidizer. The 33,000-pound X-15, including the 18,000 pounds of fuel and oxidizer that it carried aloft, had achieved Mach 3.50 with the two LR-11 engines. We figured that we could achieve our transonic speed objective by using one LR-11 to get close to Mach 2 flight in an aluminum version of the M2-F1.[2]

To get a simple weight estimation for the aluminum lifting body, I compared the wingless weights of two aircraft that had used the LR-11s earlier, the X-1 and the D-

2. Paragraph based in part on Thompson, ***At the Edge of Space***, pp. 46-47, 85, but that source gives information on the LR-11 engines that were not uprated. The powered lifting bodies used uprated engines with upwards of 8,000 lbs. of thrust versus the 6,000 lbs. of the original LR-11. See Frank Winter, "'Black Betsy': The 6000C-4 Rocket Engine, 1945-1989. Part II," ***Acta Astronautica*** 32, No. 4 (1994): 314-17, and David Baker, ***Spaceflight and Rocketry: A Chronology*** (New York: Facts on File, 1996), p. 167.

558, coming up with a target weight of about 10,000 pounds. I estimated the vehicle weight of the aluminum M2-F1 would be 5,000 pounds, including one LR-11 engine. Given the large volume inherent in the lifting-body shape, I foresaw little difficulty in installing tanks to carry another 5,000 pounds of fuel and oxidizer, bringing the launch weight up to about 10,000 pounds.

One problem arose right away in designing the vehicle. In the unpowered M2-F1, the pilot and ejection seat had been positioned on the aircraft's center of gravity, where the fuel tanks would need to be in the rocket-powered version. Fortunately, there was enough depth in the basic M2-F1 shape to move the pilot and canopy forward of the center of gravity. Earlier I had hoped that we could preserve the M2-F1's original shape in the aluminum version so that the wind-tunnel and flight data measured on the original version would remain valid for the aluminum follow-on. Moving the canopy forward, however, meant aerodynamic changes that made new wind-tunnel tests mandatory.

I calculated the aircraft's performance, assuming an air launch of a 10,000-pound M2-F1 from a B-52 at 45,000 feet, with an 8,000-pound-thrust LR-11 engine burning down to a burnout weight of 5,000 pounds. The result showed that a speed close to Mach 2 could be achieved.

Birth of the M2-F2

The cost of a rocket-powered lifting-body program could be cut substantially, I found, by using the present facilities and personnel for maintaining and operating the LR-11 engines and by using NASA's B-52 as a mothership for launching the lifting body. We could design and fabricate an adapter to be used in launching the lifting body that would attach to the B-52's wing pylon used in air-launching the X-15. When I presented my idea for the rocket-powered lifting-body program to Paul Bikle, he said it sounded good and suggested I try the idea out on others at the NASA Ames Research Center.

I presented the idea to members of the NASA Ames wind-tunnel team, including Clarence Syvertson, George Kenyon, and Jack Bronson. While they liked the idea, they said the "elephant ears" on the M2-F1 would have to go because they would burn off during re-entry from space. They were concerned that these outer horizontal elevons would create a serious heating problem from shock-wave and boundary-layer interaction as well as shock-wave impingement. I tried to talk them into leaving them on, knowing the elevons worked very well for roll control on the M2-F1 and provided a lot of roll damping to help retard any potential problems in roll oscillation. But they insisted that they had to go, saying there were no materials that could take the potential heat that would be generated on the elevons' leading edge and the slot between the elevon and vertical tail. After the NASA Ames team presented the M2-F2 configuration that it recommended for space re-entry, the team said we should use the M2-F2 in place of the M2-F1 shape in a rocket-powered transonic research program.

The roll control on the M2-F2 consisted of split upper flaps of the sort we had originally built on the M2-F1 but abandoned before we had started flight-testing. The NASA Ames team also added an extra body flap on the lower surface so that the upper split flaps and the lower body flap could be opened like feathers on a shuttlecock to give the longitudinal stability needed at transonic speeds.

Even though the extra body flap caused increased drag, the NASA Ames team members defended their decision made on the basis of wind-tunnel test results. Ken Iliff and I expressed concern about adverse yaw from the split-flap roll control. They said we could cancel it out by designing an aileron-rudder interconnect into the control system. This sounded reasonable in theory, but those flaps would complicate our lives greatly when we actually flew the M2-F2. The simplest and most straight-forward design solutions had always appealed to me, and keeping the "elephant ears" still seemed to me the simplest and most direct option.

Not giving up easily, I asked the NASA Ames engineers about the pressure on the upper-body flaps caused by the aerodynamic interaction of the rudders. They said I shouldn't worry about that, for they had prevented that problem by making the rudders operate like split flaps with outward movement only. The stationary inner surface of the vertical fin adjacent to the rudder would shield the split elevon upper flaps from rudder pressure, they claimed. They defended the feature, saying the transonic shuttlecock effect was needed in both yaw and longitudinal axes. Moving both rudders outboard, they added, provided directional stability in the transonic speed region—and added more drag, of course.

Meanwhile, Iliff and the Dryden analytical team had done a great deal of work on data bases for not only the M2-F2 but for a generic M2 vehicle, the HL-10, and an earlier version of the X-24 called the SV-5. Harriet Smith and Bertha Ryan worked with simulation programmers to develop analog engineering simulators for study of the unusual aircraft dynamics of the HL-10 and the various M2 configurations. Based on dynamic studies, the team believed that a center fin was the best solution to problems of instability that they had identified. However, the Ames team argued that the increase in dihedral effect from a center fin would make the M2-F2 much more sensitive to side gusts. Milt Thompson agreed with the Ames team that it was not a good idea to make the vehicle more sensitive to gusts.

By now, the lifting-body program was snowballing. We were getting even more input continually from NASA engineers at other sites who were experienced in aircraft and spacecraft design. I began to feel it might be time for me to back off from my simple approach and let more of these experts contribute to the program. Designing control systems for lifting bodies was going to be a major effort requiring a lot of expert help, I felt. As a result, we froze the M2-F2 configuration with the forward canopy location and the greatly modified aerodynamic controls on the aft end of the body.

Paul Bikle, Milt Thompson, and I put together a program proposal for NASA Headquarters. Because of the growing importance of our activity to the future of lifting re-entry, we suggested that two M2-F2s be built at the same time to provide us with a backup in case one vehicle was damaged and to allow us to do separate experiments simultaneously. We presented our proposal to Fred DeMerritte and his bosses at NASA Headquarters. After listening to us, DeMerritte said they'd rather we substituted the NASA Langley HL-10 for the second M2-F2.

Gene Love and the contingent from NASA Langley had made presentations to NASA Headquarters the week before we presented our proposal. Given the close proximity of the Langley Research Center in Hampton, Virginia, to NASA Headquarters in Washington, D.C., it was common for Langley representatives to be at the Washington headquarters almost daily. Unofficially, NASA Langley had always been considered the "mother" research center, and NASA Headquarters seemed to be more influenced by Langley than by any other NASA research center.

Birth of the HL-10

In 1957, while Al Eggers and his NASA Ames team were studying half-cone re-entry configurations, NASA Langley researchers were conducting broader re-entry studies, including winged and lifting-body vehicles. Hypersonic studies conducted at Langley's aerophysics division were evaluating various aerodynamic shapes. Preliminary goals at Langley in design features for a re-entry vehicle included minimization of refurbishment in time and money, fixed geometry, low deceleration loads from orbital speeds, low heating rates, ability for roll and pitch modulation, and horizontal powered landing.

According to these studies at Langley, a re-entry lifting-body vehicle with negative camber (that is, with the curved portions of wing surfaces turned upside-down) and a flat bottom might have higher trimmed lift-to-drag ratios over the angle-of-attack range than those of a blunt half-cone design. The negative-camber concept was used in 1957 in developing a vehicle—initially referred to as a Manned Lifting Re-entry Vehicle (MLRV), but now referred to simply as a lifting body—that was stable about its three axes and retained a flat lower surface for better hypersonic lifting capability. These studies at Langley found that a vehicle with an aerodynamic flap, a flat bottom, and a nose tilted up at 20 degrees would be stable about the pitch, roll, and yaw axes and trim at angles of attack up to approximately 52 degrees at a lift-to-drag ratio in excess of 0.6.

In a paper presented at the 1958 NACA Conference on High-Speed Aerodynamics, NASA Langley's John Becker described a small winged re-entry vehicle embodying all of the features that had earlier been identified as design goals at Langley, including low lift-to-drag ratio for range control, hypersonic maneuverability, and conventional glide-landing capability.[3] The vehicle in Becker's paper also

3. John V. Becker, "Preliminary Studies of Manned Satellites—Winged Configurations," *NACA Conference on High-Speed Aerodynamics: A Compilation of Papers Presented* (Moffett Field, CA: Ames Aeronautical Laboratory, 1958), pp. 45-57.

included a flat-bottomed wing with large leading-edge radius and a fuselage crossing the protected lee area atop the wing. This configuration, however, wasn't selected to carry the first American astronaut into space. Officials at Johnson Space Center opted instead for a ballistic capsule, the Mercury "man in a can." Their decision, however, did not deter researchers at NASA Langley from continuing to develop concepts and design goals for a lifting re-entry vehicle.

In the early 1960s in its space mission studies, Langley's astrophysics division began moving away from winged to lifting-body configurations. The first seven refined mission vehicle goals of 1962 echoed the desirable characteristics of re-entry vehicles described in these studies. One goal was a hypersonic lift-to-drag ratio near 1 without elevon deflection, thus avoiding heating problems near the elevons in the maximum heating portion of the trajectory. Another goal was high trimmed lift at hypersonic speeds, providing high-altitude lift modulation. A subsonic lift-to-drag ratio of approximately 4 was desirable for horizontal runway landings without power during emergencies. Furthermore, the vehicle's body should provide high volumetric efficiency (the ratio of the useful internal volume to the total exterior volume encompassed by the external skin) with a 12-person capability, and it should have acceptable heating rates and loads at all speeds, possibly including super-orbital ones. Also essential were launch-vehicle compatibility and stability and control over the speed range.

Evolving configurations at Langley were refined to meet these mission goals. Trade-off studies interrelated sweep, thickness ratio, leading-edge radius, and location of maximum thickness. The negatively cambered HL-10 lifting-body design emerged in 1962. It then entered an intermediate stage of evolution, involving nearly every research division at Langley in intensive efforts to identify and find solutions for problems associated with this type of configuration. Interestingly enough, much debate still raged over negative camber versus no camber (or symmetrical shape), fueling even more detailed studies. In the end, negatively cambered and symmetrical (no camber) configurations were evaluated in terms of the mission goals. Three more mission goals were also added, becoming serious issues in selecting camber: lower heating rates and loads comparison, lower angle of attack for a given subsonic lift-to-drag ratio, and reduced subsonic flow separation. The negatively cambered HL-10 met nine of the ten mission goals, the symmetrical design meeting only five. The only goal not met by the HL-10 was the lower angle of attack for a given subsonic lift-to-drag ratio.

The HL-10 evolved as a flat-bottomed, fixed-geometry body with rounded edges and a split trailing-edge elevon capable of symmetric upward deflection, providing the pitch trim and stability required for hypersonic re-entry and subsonic flight. The trailing-edge elevon would also deflect differentially for roll control. For even more directional stability, tip fins were added. The lower surface was negatively cambered, assuming a rocking-horse shape to provide longitudinal trim. The aft end of the upper surface was gradually tapered, or boat-tailed, reducing subsonic base drag and decreasing problems in transonic aerodynamics. There was enough forward volumetric distribution within the HL-10 to meet center-of-gravity requirements for subsystems and crew in balancing the vehicle for flight.

As research and development on the final vehicle design began at Langley, research centered on such issues as trajectory analysis and entry environment, heat transfer, structures and thermal protection, aerodynamics, dynamic stability and control, handling qualities, landing methods, emergency landings on land and water, equipment and personnel layout, and viscous effects including Mach number, Reynolds number aerodynamic scaling factor, and vehicle length.

Because the shape resembled that of a hydroplane racing boat, Langley also conducted tests with HL-10 models for horizontal landings on water, using its water test basin facility. However, even more water-landing tests would have been needed to optimize the HL-10's shape for water landings.

A disadvantage then and now of lifting bodies is that they suffer an aerodynamic heating penalty due to the fact that they spend more time within the entry trajectory than do ballistic missiles. Consequently, methods of thermal protection were extensively researched. Using small and thin-skinned inconel models, engineers also made detailed wind-tunnel tests, measuring heat-transfer distributions at Mach 8 and 20. In great detail, experimental heating was measured on the models' shapes.

The volumetric efficiency for the proposed HL-10 was relatively high in several designs. One 12-person configuration had an estimated length of 25-30 feet, a span of 21 feet, and a pressurized volume of 701 cubic feet. It also had an attached rocket adapter module and a full-length raised canopy. Some vehicle designs were 100 or more feet long.

The camber issue settled, by 1964 the HL-10 had assumed a swaybacked shape, like that of a child's rocking horse. To determine the best fin configuration, Langley conducted studies using ten wind-tunnel models—ranging from a 4.5-inch hypersonic one with twin vertical fins to a 28-foot low-speed version with a single central dorsal fin. Researchers investigated single-, twin-, and triple-fin arrangements, both lower-outboard and dorsal, along with various modifications to the aft end of the vehicle's body. Finding an acceptable fin arrangement involved a compromise between subsonic trimmed performance and hypersonic trim and stability. Langley proposed that we build the configuration that offered the best compromise, a triple-fin HL-10.

NASA-Northrop Program: Building the M2-F2 and HL-10

I formed a team at the NASA Flight Research Center that then wrote a Statement of Work for designing and fabricating the M2-F2 and HL-10. Besides furnishing the LR-11 rocket engines, NASA would provide all wind-tunnel data as well as aerodynamic load and B-52 captive-load specifications. NASA would also do all control-system analysis and simulation needed for specifying control laws and gains in the automatic functions of controls. The contractor's main responsibility would be to design and build the hardware in concert with the NASA analytical team.

Fortunately, operations engineers and technicians on the X-15 program helped us write the specifications on pilot life-support, electrical power supply, hydraulic control, landing gear, rocket, and rocket fuel subsystems. One of my long-time friends, John McTigue, then operations engineer on the third X-15, helped me specify the work for operational systems. Milt Thompson and Bruce Peterson helped me write the portions relating to the pilot's controls and cockpit displays.

In February 1964, having authorization from the NASA Associate Administrator, Raymond Bisplinghoff, we went "on the street" with a Request for Proposal (RFP), soliciting bids from 26 aerospace firms for designing and fabricating the two rocket-powered lifting bodies. Fortunately, several companies were interested in our program, believing that the next generation of spacecraft would have horizontal landing capability and that any aerospace contractor participating in our experimental lifting-body program would have an edge over other firms in later space programs. Five companies submitted bids, and our choice eventually was narrowed to two of them: North American Aviation (later to become Rockwell International, Rockwell's aeronautics and space divisions now part of Boeing) and Northrop Corporation.

Many supposed that North American (later selected as the prime contractor on the Apollo program) would be a shoo-in for the job, since North American had built the X-15. However, the Norair Division of Northrop clearly had the superior bid, the NASA Flight Research Center awarding the contract to Northrop on 2 June 1964. The RFP's timing worked in both Northrop's and NASA's favor. Northrop had intact the team that had just finished developing the prototype T-38 aircraft. A 19-month interval between the T-38 and another major program allowed Northrop to assign this team of their best people to our lifting-body program. Consequently, Northrop could keep this team together while NASA got the best bargain in skilled people for its program. Northrop's proposal listed all key persons from this team that would be working on our program, providing us as well with their resumes. Ralph C. Hakes of Northrop was assigned as Project Director with Fred R. Erb serving as Northrop's chief systems and mechanical designer.

Northrop's proposal presented a detailed preliminary design with drawings showing the use of many off-the-shelf components, including modified T-37 ejection seat, Northrop's T-38 canopy operating/locking mechanism and ejection system, T-38 stick grip, modified T-39 dual-wheel nose gear, Northrop's F-5 main gears with T-38 wheels and brakes, Northrop's X-21 hydraulic control actuators, and silver-zinc batteries for hydraulic and electrical power. Northrop signed a fixed-price contract requiring delivery of the two vehicles in 19 months for $1.2 million each, a bargain-basement price for NASA, even in the 1960s. According to one aerospace spokesman, at that time the M2-F2 and HL-10 could have cost $15 million each. In the mid-1960s, Northrop was non-union, giving the corporation flexibility in adapting the most economical and efficient methods for producing the two lifting bodies. Northrop not only delivered the vehicles on time but also did so with no cost overruns, two out-of-the-norm accomplishments for aerospace programs to that time and since.

Northrop purposely kept its project organization lean and flexible, with an average of 30 engineers and 60 shop personnel, each averaging 20 years of aerospace experience. As Ralph Hakes later recalled, the engineers involved were "all twenty-year men who had worked to government specifications all their lives and knew which ones to design to and which to skip." He added that NASA's "people and ours would talk things over and decide jointly what was reasonable compliance with the specifications. Decisions were made on the spot. It didn't require proposals and counter-proposals."[4]

NASA and Northrop's program managers devised a Joint Action Management Plan accenting five guidelines for efficiency: keep paperwork to a minimum, keep the number of employees working on the project to a minimum, have individuals—not committees—making decisions, locate the project in one area where all needed resources could be easily and quickly gathered, and fabricate the vehicles using a conservative design approach. Consequently, engineering and factory areas were located in the same building, and veteran shop technicians fabricated and assembled components from a minimum of formal drawings and—in some cases—solely from oral instructions. A special photographic process transposed drawings onto raw metal stock, avoiding costly jigs and fixtures. Northrop's project personnel maintained a very close operational relationship with NASA's personnel, maximizing the joint team's ability to react swiftly in solving problems and making changes.

The overall tone of cooperation in this joint NASA-Northrop program had been established from the beginning by Paul Bikle and Northrop's Richard Horner. The two men had much respect for each other and a good person-to-person understanding of how the program was to be conducted. Horner and Bikle had worked together often in the past. Horner had worked for the Air Force from 1945 until June 1959, when he became NASA associate administrator until July 1960. Afterwards, he became executive vice president of Northrop. Together, Bikle and Horner agreed to do away with red tape and unnecessary paperwork, a simplification that had a dramatic effect on keeping costs low and efficiency high. Both men had impeccable reputations and credibility, keeping their word on agreements. Even though this was a fixed-price contract, Bikle and Horner agreed that it would be to both NASA's and Northrop's best interests to build these lifting bodies in the most cost-effective and timely manner.

The Program That Almost Was: Little Joes and the M2

About this time, another opportunity arose to conduct a low-cost program using surplus equipment. Four Little Joe solid rockets, used to test the Apollo capsule's escape system, were available at the NASA White Sands rocket testing facility in New

4. Quoted in Hallion, *On the Frontier*, p. 154.

Mexico. I began to explore the possibility of mounting an M2-F2 configuration on top of a Little Joe booster for a vertical launch and possible flight to Mach 6.

Earlier, before it was assembled with the steel-tube carriage structure, I had had a fiberglass mold made from the M2-F1 wooden shell, just in case it was damaged in flight tests or we wanted to build another, heavier, M2 out of fiberglass instead of wood. Using this mold, I could make a vehicle with a thick fiberglass skin capable of withstanding speeds up to Mach 6. The vehicle would be made much like a boat, its thick skin acting like an ablative coating to cool the structure from aerodynamic heating at high speeds. It would be unpiloted with a rocket climb and push-over trajectory followed by a pre-programmed turn.

Also available for recovering the M2 after it had slowed down to about Mach 2 were some surplus parachute systems from the Gemini program. John Kiker of the Johnson Space Center, in charge of developing the parachute spacecraft recovery systems for NASA's Gemini and Apollo programs, offered his services in adapting the parachute systems for recovery of the M2. Once we found out we had mutual interests in flying experimental radio-controlled model airplanes, Kiker and I became and have remained friends. In the early days of Shuttle development, Kiker had constructed flying scale models of the Boeing four-engine 747 and the ***Enterprise***, then demonstrated a successful launch of the model ***Enterprise*** from the back of the model 747 at Johnson Space Center. This test, using Kiker's models, was done before the approach-and-landing tests of the full-scale ***Enterprise*** at Edwards AFB in October 1977.

After I talked with Kiker, I telephoned Dick Thompson, the manager of the NASA White Sands facility, about using the Little Joe boosters to launch an M2. Thompson liked the idea and said he could furnish the personnel for servicing the rockets, preparing them for launch, and conducting the launch operation, if the Flight Research Center would be responsible for the M2 payload. I found myself trying hard to restrain my excitement, for I had already located a surplus hydraulic control system and a programmable missile guidance system. It was all going too smoothly, too quickly, too easily to be believable. About then, a big dose of reality intruded, ending this tiny program before it had even begun.

Dick Thompson contacted me, saying the Little Joe rockets were out-of-date and would require an inspection before they could be used. Being naive about how much such things cost, it didn't occur to me that it would cost very much to inspect something as simple as a solid rocket. So it blew my mind when Thompson told me that an inspection would cost about $1,000,000 per rocket—about half the cost of the Northrop contract for the two lifting bodies. Apparently, the inspection involved much more than simply x-raying the solid propellant for cracks.

I reasoned with Thompson, trying to find a way to use an abbreviated inspection since the test flight would be unpiloted. Thompson was adamant, however, opposed to allowing even the potential for an explosion on the launch pad, NASA space policy having become very conservative after the early days of numerous rocket explosions on the pad.

Thus ended the program that almost was. It had been a good idea, just not a practical one. In the future, others' ideas would have better chances for success.

NASA-Air Force Lifting-Body Program

Since 1960, the Air Force had also been conducting studies of piloted, maneuverable lifting-body spacecraft as alternatives to the ballistic orbital re-entry concepts then in favor. Given the long history of cooperation and joint ventures between the Air Force and the NASA Flight Research Center at Edwards AFB, it was only natural for them eventually to pool their resources in the flight-test portion of the lifting-body program, much the way they had in the X-15 and earlier X-plane programs.

Much as Walt Williams had done before him, Paul Bikle had always worked closely and effectively with others at the Air Force Flight Test Center. This spirit of cooperation extended to all personnel levels. Since the early days of the NACA station at Muroc in the late 1940s, there had been few, if any, disagreements at the work-level between NASA and Air Force personnel, and any that existed had been imposed from above.

In the early spring of 1965, as Northrop entered its final months of fabricating the first of the two heavyweights, Paul Bikle recognized that the lifting-body program was, like the X-15 program before it, becoming too large for the Flight Research Center (FRC) to manage and operate alone and that NASA and the Air Force had similar interests in the lifting bodies. Bikle met with his Air Force counterpart, Major General Irving Branch, commander of the Air Force Flight Test Center (AFFTC), throughout the early spring.

From those meetings emerged a memorandum of understanding between the two centers on 19 April 1965, nearly two months before the M2-F2 was completed at Northrop's plant in Hawthorne, California. Drawing on the two centers' shared experience with the X-15 program and alluding not only to the excellent working relationship between NASA and the Air Force but also to similarities between the X-15 and lifting-body programs, the memorandum of understanding created the Joint FRC/AFFTC Lifting-Body Flight Test Committee. Ten members made up the committee headed up by Bikle as chairman and Branch as vice-chairman. Six of the remaining eight members included one representative each from the NASA and Air Force pilots, engineers, and project officers. A NASA instrumentation representative and an Air Force medical officer completed the committee.

The joint flight-test committee had responsibility not only for the test program but also for all outside relations and contacts. Maintenance, instrumentation, and ground support for the vehicles remained the responsibility of the Flight Research Center. The Air Force Flight Test Center assumed responsibility for the launch and support aircraft, the rocket power plant, the personal equipment of the pilots, and medical support. The two centers assumed joint responsibility for research flight planning, flight data analysis, test piloting, range support, and overall flight operations.

John McTigue, NASA lifting body project manager. (NASA photo EC76 5352)

Career Decision: Manager or Engineer?

As the lifting-body program grew larger, it needed a full-time manager to coordinate activities among the NASA Flight Research Center, the Air Force Flight Test Center, and the contractor, Northrop. Called into Paul Bikle's office one day, I was confronted with a career decision. Bikle gave me a choice. I could move into management of the program, which would pull me away from involvement in day-to-day technical and engineering activities, or I could stay with engineering.

Bikle told me that he thought I would be happier and NASA would benefit more if I remained in engineering, free to continue generating new technical ideas. He said that if I continued working within the program's technical engineering team, I could serve as NASA's project engineer, coordinating all technical activities for the NASA and Northrop engineering teams. He gave me a few days to make my decision, saying that if I opted to remain in engineering, he would appoint John McTigue as lifting-body project manager. I decided to stay with engineering.

As an operations engineer on the X-15, McTigue had gained experience in scheduling crews and technicians to meet flight schedules. Bikle believed McTigue would make good use of this experience in building up and servicing all systems needed to operate the lifting bodies. McTigue was also very familiar with the rocket, hydraulic, and life-support systems of the X-15s, which were, in most cases, identical to those of the lifting bodies. Furthermore, Bikle earlier had created a competitive spirit among the three X-15 operations engineers in meeting or beating flight schedules by betting against these engineers. On several occasions, Bikle had lost his bet and McTigue had won. Obviously, Bikle was impressed with McTigue as a manager who would keep the program on schedule.

NASA and Northrop Single-Team Engineering

Having a flight-test facility at Edwards AFB for testing F-5s, T-38s, and X-21s, Northrop ran a little commuter-plane operation daily between its plant in Hawthorne in the Los Angeles basin and Edwards AFB, using a couple of Piaggio twin-engined airplanes. During the 19 months of the lifting-body contract, I and the rest of the NASA engineering team commuted almost daily by Northrop's planes to the Hawthorne plant, where I spent nearly half of my time during this period.

In the lifting-body program, NASA engineers did not have the "do-as-I-say" relationship with Northrop engineers that was typical between customer and contractor in the aerospace industry. Instead, we worked together as a single team to make the best possible product. The keys to our success were mutual respect, trust, and cooperation. The Northrop engineers respected and trusted not only the expertise of the NASA engineers in aerodynamics and in stability and control analysis but also our operational experience with rocket-powered aircraft. Equally, the NASA engineers trusted and respected the outstanding ability of the Northrop engineers in fabricating airframes. Working one-on-one in small groups, we made on-the-spot decisions, avoiding the usual time-consuming process of written proposals and counterproposals in solving problems and making changes.

One day, we were called together by a Northrop engineer named Stevenson who was responsible for the M2-F2's weight and balance. He showed us through his latest calculations that maintaining weight balance on the M2-F2 was becoming a large problem, given the twin challenges of a narrow nose area, limiting space for systems, and the requirement for locating multiple actuator systems in the aft end with all control surfaces. We needed to do something drastic to restore balance by putting ballast, or weight, in the nose. Otherwise, the vehicle would be tail-heavy.

An aircraft designer usually considers having to add ballast to a new vehicle as a negative reflection on his or her ability to provide an efficient design. Ballast adds nothing desirable. It puts higher loads on the structure and decreases the aircraft's performance.

We faced a large dilemma. The usual solution would have been to put depleted uranium around the pilot's feet in the aircraft's nose. Having a much higher density than lead, depleted uranium is commonly used for balance in aerospace vehicles when there is limited room for ballast. However, NASA pilots Milt Thompson and Bruce Peterson, as well as the chief Air Force lifting-body pilot, Jerry Gentry, didn't like the idea of cooking their feet in radiation, so we had to come up with another solution.

Stevenson did a cost trade-off study for using gold as ballast in the nose. He also demonstrated how the high-density gold bricks could be cut and fitted into the structure around the pilot's feet without blocking the pilot's vision through the nose window. The $35-per-ounce price for gold at the time was still cheaper than the labor costs would be for balancing the vehicle by redesigning the structure in the aft portion of the M2-F2 and moving equipment forward.

The little group of NASA and Northrop engineers sat around a table, equally desperate to solve this problem. By the time this problem arose, the two teams of engineers had coalesced into one. Everyone focused on solving the problem, not pointing fingers at others' mistakes. Thinking aloud, I suggested that if we could actually put something useful in the nose, rather than simply adding ballast, we might salvage our pride as designers. Immediately, another engineer suggested we put some extra structure around the pilot to give him added protection in case of a crash. As a group, we jumped on that idea, with no debate or dissent, and within thirty minutes we had solved the problem by changing the design, replacing the 50G cockpit with a nearly 300G cockpit that had a very heavy steel frame around the pilot. As it turns out, the decision to add the protective cage-like structure around the pilot helped to save pilot Bruce Peterson's life when the M2-F2 crashed two years later. Only the cockpit remained intact in that horrendous accident that left the rest of the aircraft looking like a crumpled beer can at a Hell's Angels' party.

In similar ways, we approached and solved other engineering problems as they arose. Time used for casting blame and engaging in agonizing debates over proposed solutions simply leaves that much less time for designing and building. Furthermore, the NASA engineers mainly considered themselves to be support and backup for the Northrop team working at the Hawthorne plant. For example, rather than asking the Northrop team to come to Edwards AFB for meetings or for looking at hardware, we would hold the meetings or take the hardware to Hawthorne, minimizing loss in time.

Things were going so smoothly, unlike typical aerospace projects, that something just had to happen—and it did.

NASA Langley Modifies the HL-10

After the contract had been signed by Northrop and the Flight Research Center, NASA Langley continued wind-tunnel tests on the HL-10 and discovered that the trimmed subsonic lift-to-drag ratio was only slightly more than 3, considerably below Langley's established goal of 4. Furthermore, negative directional stability showed up at low supersonic speeds and at some angles of attack.

To fix these problems, Langley initially considered adding an ejectable tip-fin scheme, only to discard the idea, finding it unacceptable to be ejecting tip fins during the final phase of a mission. Then, working from wind-tunnel test results, Langley engineers changed the tip-fin shape, developing a configuration that increased area, toe-in angle, and roll-out angle. They also added simple two-position flaps to the trailing edge of the tip fins and upper elevon to vary the base area. Closing these flaps would also minimize the subsonic base drag. This modification brought the maximum lift-to-drag ratio to nearly 3.4, still short of the target 4.0. However, it improved the directional stability.

On 3 February 1965, nearly 10 months into the 19-month contract with Northrop, Langley presented its proposed HL-10 modification at a meeting held at the Flight Research Center. Attending the meeting were several of the top Langley engineers—including Eugene Love, Robert Rainey, and Jack Paulson—as well as NASA Headquarters' Fred DeMerritte, chief of the lifting-body program for the Office of Advanced Research and Technology, through whom we received our funding for the follow-on lifting-body program. The proposal was to add six more control surfaces to the HL-10. These would be two-position surfaces consisting of elevator flaps, located on the upper surface of the elevon, and outboard tip-fin flaps.

The result was a required design change and modification to the contractual agreement with Northrop. The modification was done as required, but it was done minus the wholehearted support of NASA and Northrop program managers and engineers. However, later in the HL-10 program, the required change came to be seen as an excellent decision. The modification simplified the flight-control design. It also allowed the pilot to move during flight from subsonic to supersonic speeds simply by throwing a switch, requiring less trim change in the pilot's control-stick position. The pilot could now easily convert the HL-10 from a "shuttlecock" to a low-drag subsonic configuration.

Included in the modification was an enlargement of the center and tip fins that improved trim and stability at transonic and supersonic speeds and increased the lift-to-drag ratio in the approach to landing. At subsonic speeds and during landing, the two-position flaps on the upper elevon surface, split rudder, and tip fins retracted for maximum boat-tailing (minimum base area) on the aft portion of the vehicle. At high subsonic, transonic, and supersonic speeds, the movable flaps deflected outwardly, minimizing flow separation at control surface areas.

M2-F2 Roll-Out

The modification to the HL-10 meant that the M2-F2 was the first to be finished, rolling out of Northrop's Hawthorne plant on 15 June 1965. The next day, it was trucked over the mountains north of Los Angeles to Edwards AFB. At its unveiling, the M2-F2 lacked the LR-11 rocket engine, but we planned to fly it first as a glider, then modify it for powered flight.

Made of aluminum, the M2-F2 weighed 4,630 pounds, was 22 feet long, and had a span of 9.4 feet. Its retractable landing gear used high-pressure nitrogen to extend the landing gear just before touchdown. The boosted hydraulic control system was pressurized by electric pumps running off a bank of nickel-silver batteries. A Stability Augmentation System (SAS) in all three axes helped the control system in damping out undesirable vehicle motions. For instant lift to overcome drag momentarily during the prelanding flare, the pilot could use the vehicle's four throttleable hydrogen-peroxide rockets, rated at 400 pounds each. The M2-F2 also had a zero-zero seat, a modification by Weber of the one used in the F-106 Delta Dart.

We put the M2-F2 next to the M2-F1 for a family photograph. Except for being identical in size, there were few similarities. The M2-F2 lacked the M2-F1's "elephant ears," had an extended boat-tail and forward canopy, and would eventually weigh 10 times as much as the M2-F1.

M2-F2 Wind-Tunnel Tests

Soon after the first heavyweight lifting body arrived at the NASA Flight Research Center, more team members were assigned to the M2-F2, including operations engineer Meryl DeGeer, crew chief Bill LePage, and assistant crew chief Jay L. King. In helping to ready the M2-F2 for flight testing and research, Bill Clifton did the instrumentation engineering and John M. Bruno, Al Grieshaber, and Bob Veith installed the flight research instrumentation.

Since full-scale testing of the M2-F1 had worked out well, the NASA Ames wind-tunnel team suggested that we measure the M2-F2's aerodynamic characteristics at landing speeds in the 40-by-80 wind tunnel. DeGeer and LePage agreed, wanting to test under wind-tunnel conditions the vehicle's control system, landing-gear deployment, and emergency ram-air turbine that would provide hydraulic power for operating the controls if the battery driving the pumps failed in flight. By August 1965, 100 hours of wind-tunnel tests would be completed on the M2-F2 within a period of two weeks.

In late July, the M2-F2 was loaded on a truck for its trip north to the NASA Ames Research Center, stirring up memories for many of us of the similar trek two years earlier with the M2-F1. This time, however, the wind-tunnel testing would be more complex than that done on the "flying bathtub." Several changes replaced nearly everything done by the person who had sat in the cockpit throughout the M2-F1's tests—as well as much of the hand-plotting of test data—allowing the wind-tunnel tests on the M2-F2 to move along more rapidly.

Hoses ran from an aircraft hydraulic power cart to the vehicle atop the pedestal, powering its control system. Pilot linkages from the cockpit to the hydraulic servos were replaced with miniature electric screw jack actuators. Toggle switches in the wind-tunnel's control room activated these actuators that, in turn, controlled the hydraulic actuators moving the control surfaces to various settings.

We also made use of the flight instrumentation onboard the M2-F2, parking one of our mobile ground-receiving stations outside the wind tunnel and hard-wiring it to the vehicle's instrumentation. In this way, sensors inside the aircraft allowed air speed, angle of attack and sideslip, and control positions to be recorded along with data from the wind tunnel's measuring system. With all this help replacing what had earlier been done only by human hand during the M2-F1's tests, Bertha Ryan could assume sole responsibility for assimilating all wind-tunnel data on the M2-F2. Nevertheless, there still remained a lot of data-plotting that had to be done by hand.

We began with testing the operational systems, which required a person in the cockpit to operate the landing-gear deployment handle and the ram-air turbine unit. DeGeer volunteered and climbed into the cockpit. However, the vehicle's canopy had been covered with paper to protect it from scratches during the tests, and DeGeer began to get claustrophobic right away. LePage opened a peephole in the paper so DeGeer could see outside.

After the wind tunnel was brought up to speed, it began to get hot in the cockpit, seemingly due to all the bright lights used to illuminate the vehicle. Trying to cool the interior of the cockpit, DeGeer opened the ram-air doors. Of course, the air coming into the cockpit was even hotter, the tunnel actually heating the air. Despite his discomfort, DeGeer deployed the landing gear and the ram-air turbine. Both systems worked well, and we could move along to the aerodynamic testing that didn't require literally having a warm body in the cockpit. Afterwards, DeGeer said he had gained great appreciation from his own experience for what Dick Eldredge must have endured two years earlier, sitting in the cockpit of the M2-F1 in the wind tunnel for eight hours.

Wind-Tunnel Tests of M2-F2, HL-10, and B-52 Models

Because of the potential for either heavyweight lifting body to collide with the B-52 motherplane immediately following launch, we conducted another set of wind-tunnel tests in 1965, this time at NASA Langley, using models of the B-52 bomber and the M2-F2 and HL-10 lifting bodies. During these tests, the airflow around the lifting body hanging in launch position was deflected upstream by the B-52's nose as well as (near and just above the lifting body) by the B-52's wing. This indicated that angular flow could cause the lifting body to roll and pitch immediately following hook release from the B-52. Since this could occur in a mere fraction of a second, the pilot would not be able to react fast enough to avoid a roll-off and possible vertical-fin contact with the B-52's launch pylon. In some cases, the automatic and gyro-driven rate damper might be able to react that quickly, the controls preset before launch to counter any unwanted motions after launch, but it was just as likely to be too slow to keep the lifting body from making contact with the B-52.

Launch studies by Wen Painter and Berwin Kock found that the M2-F2's vertical fins would make contact with the B-52's pylon used in launching the X-15. Consequently, the adapter used for launching the M2-F2 from the pylon was modified to lower the lifting body. In the wind tunnel, the lifting-body model was positioned at different points below the B-52 as well as in launch position, with forces and moments measured on the M2-F2 then used to calculate the vehicle's flight path and attitude as it fell away from the B-52. Similar wind-tunnel tests much earlier on a model of the X-15 had also succeeded in predicting the motions of the X-15 after launch from the B-52. Our tests used the same B-52 model that had been used in the X-15 wind-tunnel tests.

Years later, Jerry Gentry, one of only four pilots to fly the M2-F2, recalled how he and others downplayed the fear that still existed after the wind-tunnel tests that the lifting body might fly back up into the B-52 after it separated from the pylon. "There was no question which way you were going when the B-52 dropped you," he said. "One guy used to say that if they dropped a brick out of the B-52 at the same time [he] released, [he]'d beat the brick to the ground."[5]

Moving Toward Flight

After we trucked the M2-F2 back to Edwards AFB, we began preparing for its first glide tests. Our staff expanded to meet these needs. Added to assist DeGeer were Norm DeMar, who acted as lead systems engineer, and Northrop's Jim Crosby, systems electrical engineer for the yet-to-be-installed rocket engine. The crew under the direc-

5. Wilkinson, "Legacy of the Lifting Body," p. 55.

Cornell T33A, equipped with a computer control system to simulate predicted flying qualities of the M2-F2. This aircraft was later outfitted with drag devices to simulate the steep glide slope of the M2-F2. (NASA photo EC87 126-7)

tion of crew chief LePage and assistant crew chief King grew to include mechanics Chet Bergner and Orion Billeter, electrician and electronic technician Millard I. Lockwood, and inspectors Bill Link and John E. Reeves. For seven months, Jack Cates, Mil Lockwood, and Wen Painter worked on the problems remaining in the Stability Augmentation System, resolving them by May 1966.

As with the M2-F1, Milt Thompson was selected by Bikle and Chief of Flight Operations Joe Vensel to pilot the M2-F2 in its first glide test. A list of five more future pilots for the M2-F2 was also drawn up, including NASA pilots Bruce Peterson, Bill Dana, and Fred Haise as well as Air Force pilots Donald Sorlie and Jerry Gentry. As the "angry" qualities of the M2-F2 revealed themselves later in actual flight, only three of these pilots—Peterson, Sorlie, and Gentry—would, in addition to Thompson, actually get to fly the M2-F2.

As part of the pilot preparation for the first flights of the M2-F2, Ken Iliff and Larry Taylor designed a flight experiment, using a highly modified and variable-stability Lockheed T-33A jet trainer from the Cornell Aeronautical Laboratory of Buffalo, New York, to simulate the flight characteristics of the M2-F2. When the petal-shaped surfaces called "drag petals" that had been installed on the T-33A's wing-tip tanks were extended in flight, the aircraft's lift-to-drag ratio varied from its usual 12-14 to as low as 2, approximating the lift-to-drag ratio of the M2-F2. The T-33A was part of a cooperative pilot training and aircraft simulation program that the NASA Flight Research Center had launched earlier with Cornell, the T-33A used initially to simu-

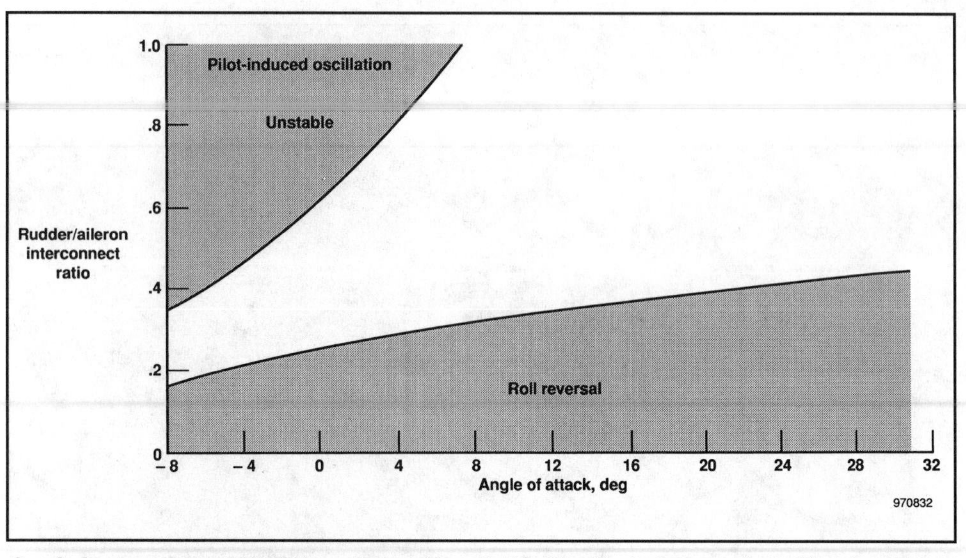

Graph showing predicted M2-F2 lateral control boundaries. The pilot had to reset the rudder/aileron-interconnect control wheel in the cockpit as the angle of attack changed in order to avoid loss of lateral control.

late the low lift-to-drag ratio characteristic of the X-15 during re-entry. The T-33A was used at the Flight Research Center for in-flight simulation of the M2-F2 in the winter and spring of 1965, with Cornell's test pilot Robert Harper and then Thompson, Peterson, and Haise executing typical lifting-body approaches in the T-33A.

The analytical team consisting of Iliff, Bertha Ryan, Harriet Smith, and others was concerned with the results of this flight experiment as pinpointing a potential lateral control problem in the M2-F2, although the pilots felt they could live with the problem after flying the T-33. In any event, they were well aware, before the actual M2-F2 flights began in July 1966, that lateral control of the craft would require considerable attention and technique on their parts. Iliff suggested delaying the flight tests until a center fin or a control scheme with a lead-lag compensator could provide a suitable fix to the lateral control problem. However, Milt Thompson, with backing from the Ames wind tunnel team, believed the problem could be solved with proper control rigging and pilot technique.

I then went along with Thompson and the Ames team, but it bothered me that Iliff in particular was not happy with the approach. We had Northrop install a small wheel in the left side of the cockpit so the pilot could adjust the rudder aileron interconnect in flight. Thompson continued to express confidence that the pilots could rely on their skills to adjust the rudder aileron interconnect ratio manually in flight.

The interconnect ratio had to be high to roll the M2-F2, due to its extremely high dihedral at high angles of attack as well as adverse yaw of the differential upper flaps (elevons). At low angles of attack and high speed, however, using too much rudder for roll control would result in a pilot-induced oscillation. If the pilot did not set the interconnect wheel properly to match flight conditions, he could have serious problems controlling the vehicle in roll. Indeed, we were asking a lot from the M2-F2 pilots.

Little did we know then that in the M2-F2 we had created a monster ready to bite the first time a test pilot became distracted.

We weren't in a rush to make the first glide flight, preferring to be absolutely sure that everything was in order. We did seven captive flights with Milt Thompson sitting in the M2-F2 attached to the B-52's X-15 pylon. Operational anomalies turned up on each of the captive flights that had to be corrected on the flight that followed. The captive flights turned out to be excellent rehearsals for everyone involved in the control room, on the ramp, in the B-52, and, of course, in the cockpit of the M2-F2.

CHAPTER 5

ANGRY MACHINES

By 1966, the Air Force was considering developing its own lifting-body configuration to add to the program. To gain experience in engineering and flight planning useful later in developing and testing its own lifting body, the Air Force participated in the M2-F2 project. Heading up the Air Force's lifting-body effort was program manager Robert G. "Bob" Hoey, who had extensive experience with the X-15 and experimental flight testing. Air Force Captain John Durrett assisted with general engineering. In January 1970, after the X-15 program ended, program engineer Johnny Armstrong joined the Air Force's lifting-body team. Although the team was relatively young, it had considerable experience in experimental flight testing.

Hoey and Armstrong had worked together as Air Force flight-test engineers in the highly successful X-15 program. Before he became NASA director at the Flight Research Center in 1959, Paul Bikle had served as technical director for the Air Force Flight Test Center at Edwards AFB. Hoey, who had been at Edwards approximately twelve years, had a good relationship with NASA management, including Bikle. The success of the X-15 program made it easy for us at NASA to consider the Air Force's lifting-body team as "the experts." Bertha Ryan worked closely with Hoey and the rest of his team as the NASA stability and control engineer and aerodynamicist for the M2-F2. Excellent communication existed between the lifting-body teams, with the Air Force offices only about a mile down the road from those of NASA.

Hoey and his team modified an X-15 simulator to use for training pilots and planning the first 15 flights of the M2-F2, while we at NASA were upgrading our own M2-F2 simulator and changing computers. Hoey's team loaded its simulator with the M2-F2 data from the wind-tunnel tests. Before the first flight of the M2-F2, Milt Thompson spent many hours on the simulator, becoming well acquainted with the vehicle's stability limits, including the boundaries for pilot-induced oscillation (PIO) and roll-control reversal.

First Flight of the M2-F2

For its first glide flight on 12 July 1966, the M2-F2 was mated with the B-52 mothership, carried aloft, then launched on a north heading at 45,000 feet. The launch was very mild, Milt Thompson reported, with at most 28 degrees of right roll following launch. The flight plan called for two 90-degree turns to the left with a landing to the south on the lakebed's Runway 18. He made a simulated landing starting at 22,000 feet, coming level at 19,000 feet between the two 90-degree turns, firing the peroxide rocket during the landing simulation with no noticeable changes in attitude (orientation) with thrust.

M2-F2 mated with B-52 to be carried aloft for launch. (NASA photo E65 13865)

Using the manual control to lower the interconnect ratio between the ailerons and rudder to 0.4 on the pushover at altitude, Milt felt that the vehicle's roll response was not great enough as he tried to begin the second 90-degree turn as planned at 16,000 feet and 190 knots. He increased the interconnect ratio to 0.6, in effect adding rudder as he began the final turn. During the turn's pushover, the M2-F2 developed an uncomfortable lateral-directional oscillation.

Milt tried to turn the interconnect ratio down, but, as he later said, he turned it the wrong way just as he was turning final. Rather than decreasing it, he had accidentally increased it to 1.25. The oscillations increased to 90 degrees, the flight films showing the vehicle swinging madly from side to side. The view through the windshield inside the M2-F2, as captured on film by the camera behind Milt in the cockpit, showed a horizon rolling rapidly from vertical to vertical. Quickly realizing the error, Milt reduced the interconnect ratio back to 0.4, which decreased rudder. He took his hand off the control stick, and the oscillations damped out rapidly.

He reached a pre-flare speed of 280 knots at 1,200 feet altitude. At flare completion, speed was 240 knots. Landing gear was deployed at 218 knots, accompanied by mild pitch transient, or change in attitude. Milt landed the M2-F2, the vehicle touching down at the exact spot planned at 164 knots, then coasting 1.5 miles across the lakebed. Lasting not quite four minutes, the first flight of the M2-F2 appeared to be an unqualified success.

Jay L. King, Joseph D. Huxman, and Orion B. Billeter assist Milt Thompson in boarding the M2-F2 attached to the B-52. (NASA photo EC66 1154)

During the debriefing afterwards, Milt apologized for nearly losing control of the vehicle by moving the interconnect wheel the wrong way. Later, we found two errors had been made in the simulator. First, by employing the Air Force's X-15 simulator cockpit, we had used the existing X-15's speed brake handle instead of the M2-F2 pilot's interconnect wheel. Second, the interconnect control direction was the reverse of the wheel in the actual aircraft. In short, Milt had been practicing with a simulator that did not represent the M2-F2, a serious foul-up that both we and Bob Hoey's Air Force simulator team found embarrassing. What might have been a disaster in the air was averted by Milt's quick adaptability and knowledge of the lifting body's characteristics. Realizing that the interconnect settings were incorrect, he took appropriate if intuitive corrective action.

One more error—this time, a minor one—was made during the M2-F2's first flight. Vic Horton had been onboard the B-52, his only task to turn on the 16mm camera 10 seconds before launch to film the top of the M2-F2 as it fell away from the B-52. He forgot to turn on the camera. After the crew briefing for the second M2-F2 flight, Wen Painter and Berwin Kock presented Horton with a "Launch Panel Camera Switch Simulator." It was made out of a cardboard box and had a large lever marked CAMERA ON/OFF. As the crew laughed, Horton turned the lever to CAMERA ON. A banana rolled out. The crew howled with laughter. Horton grabbed the banana and threw it at Painter and Kock.

Milt's Last Lifting-Body Flight

On 2 September 1966, Milt Thompson made his fifth flight on the M2-F2, his last lifting-body flight. He had decided to make a career change, moving into management with the Flight Research Center. NASA lost a superb research pilot when Milt Thompson retired from the cockpit, but we later reaped great benefit from his experience when he became chief of the research projects office at the Center in January 1967, responsible for all flight projects, including those of the X-15 and the lifting bodies. Milt never spoke publicly in those days about why he made the career change. Some surmised that he might have felt he had used up his "nine lives" in the close calls he had had as a pilot.

One of those close calls happened on 20 December 1962, about eight months before he flew the M2-F1. He was flying an F-104 chase aircraft. As he prepared for landing, he was lowering the flaps when the mechanical cross link between the right and left flaps broke. The flaps were stuck, one up and one down, and the F-104 started rolling. Somehow Milt managed to maintain altitude while the aircraft made a series of 360-degree rolls across the sky. He tried recycling the flaps and resetting the circuit breakers during the rolls, but to no avail.

As the F-104 continued to roll, Milt managed to steer it over the bombing range at Edwards AFB. Since the aircraft was only about 5,000 feet above the ground, Milt made a carefully timed ejection when the cockpit was pointed upward. He floated down in his parachute, landing safely on the bombing range.

The F-104 went down about two miles from where Milt had landed, the aircraft digging a huge black hole in the ground upon impact. Milt gathered up his parachute and walked half a mile along the edge of the bombing range to a road leading to the rocket test site on Leuhman Ridge. He stuck out his thumb and hitched a ride in a pickup truck that brought him back to the NASA building.

When he walked into the pilot's office, a full-scale search was already underway. Helicopters were landing at the crash site. No one had seen Milt eject or spotted his parachute descending. The assumption was that his body would be found in the wreckage of the F-104. The mood changed from heavy sadness to surprised relief when Milt walked into the office.

After he retired as a NASA pilot in 1966, Milt later made (to my knowledge) only one public statement about his career change, and he made it in his book, ***At the Edge of Space***, published in 1992. There, he explains it was boredom, not fear, that led to his career change, saying that he had made up his mind and even discussed the career change with Bikle nearly two months earlier, before he began flying the M2-F2 in July.

"I felt that the exciting programs were winding down," he wrote, "and I could not see any new challenging programs coming up in the near future. I really enjoyed the challenge of an X-15 flight or a lifting-body flight, but I was getting bored with

M2-F2 in landing flare, gear up, closely followed by an F-104 chase airplane. (NASA photo EC66 1567) p. 67

the routine proficiency flying that was required between research flights. When a pilot gets bored with flying, it is time to quit."[1]

Gentry Fast Forwards

By 12 October 1966, the M2-F2 had been flown ten times—five by Milt Thompson, two by NASA research pilot Bruce Peterson, and three by Air Force test pilot Don Sorlie. Sorlie also got into a PIO problem on his first flight in the M2-F2, but he had planned ahead of time what he would do if it happened, and he had sufficient altitude to execute a full recovery. After two more flights with no additional problems, Sorlie gave the okay for Air Force research pilot Jerry Gentry to fly the M2-F2.

Gentry's first flight in the M2-F2 on 12 October went smoothly according to flight plan from B-52 launch to just before touchdown. Then, the unexpected happened. At about 100 feet above the ground, mere seconds before touchdown, Gentry reached for the landing gear handle—and couldn't reach it. What happened next was the result of quick thinking. Within no more than five or six seconds, he loosened the shoulder harness, leaned forward, pulled the handle, tightened the shoulder harness, and continued with the landing.

1. Thompson, *At the Edge of Space*, p. 276.

For a second time, the M2-F2 was saved from disaster by the quick thinking and skill of the pilot. Northrop had designed the cockpit dimensions to accommodate Milt Thompson and Bruce Peterson. No consideration had been given to the needs of smaller or shorter pilots, including arm span. A second error was a faulty preflight checkout procedure, for Gentry's inability to reach and pull the landing gear handle while secured in the shoulder harness should have been discovered then, not seconds before touchdown.

Gentry became the Air Force's chief lifting-body pilot on the M2-F2 and, later, the HL-10. With the retirement of Milt Thompson from research flying, there were now only two official lifting-body pilots, Gentry for the Air Force and Bruce Peterson for NASA. Before the first flight of the HL-10 in late December 1966, Peterson made two unpowered flights in the M2-F2. Between July and late December, four pilots—Milt Thompson, Bruce Peterson, Don Sorlie, and Jerry Gentry—had made a total of fourteen flights in the M2-F2.

Air Force/NASA Simulators

When the HL-10 arrived from Northrop, it was trucked to NASA Ames for wind-tunnel testing, as had been done with the M2-F2. The only difference was that data handling was even more automated with the HL-10 than it had been with the M2-F2, thanks to our and the wind-tunnel crew's greater experience and practice in testing the earlier lifting bodies. The HL-10 project was also better staffed with NASA personnel than the M2-F2 had been, the average flight-test experience being three to six years. However, while the M2-F2 team was made up of both NASA and Air Force research or analytical engineers, the HL-10 project was essentially a solo in engineering by NASA.

Bob Hoey wanted to maintain hands-on experience with the aerodynamics of the M2-F2, even after we had developed our own M2-F2 simulator, so he decided to keep the original M2-F2 simulation at the Air Force Flight Test Center. Later, the NASA team at the Flight Test Center concentrated mainly on the simulation of the HL-10.

For a period of time, there were two M2-F2 simulators, one at the Air Force and one at NASA. Even though both simulators used the same wind-tunnel data, the way in which the data was processed and interpreted by the computers within the simulators was different. Once a week, I compared the technical results from both simulators. Generally, the simulators gave the same results. However, now and then, slight differences would appear in the results, followed by lively discussions of which were correct. I felt this was a healthy activity, especially when both simulations concluded that the M2-F2 was safe to fly and when neither set of results required alteration in the vehicle's control settings, stability augmentation system gains, or flight procedures. Joe Weil, my boss and head of NASA's research division, felt uneasy about the lively discussions, seeing them as discord. He basically felt that if there was only one

Robert G. (Bob) Hoey, Air Force lifting-body program manager. (Air Force photo, available as NASA photo EC97 44183-5)

M2-F2 simulator, the Air Force and NASA lifting-body teams would work together even more harmoniously.

New Lifting-Body Project Engineer

By 1966, I was finding my job as lifting-body project engineer more a job of managing people and solving their problems than of directing a technical effort. Once again, as I had in 1965, I found myself facing a career decision.

Over the years, I had worked with Garrison "Gary" Layton on several NASA programs. On our own time, we also had helped one another in our common hobby, flying experimental radio-controlled model airplanes. As I grew more concerned at how far I was getting away from technical engineering and into management, Gary Layton mentioned that he would like the opportunity to take over as the lifting-body project engineer so that I could have the opportunity to get back to the type of work I loved, especially developing some ideas I was having for remotely controlled vehicles.

Layton and I went to our bosses, Paul Bikle and Joe Weil, to get their approval for Layton to take over as lifting-body project engineer. Once the change was approved in 1967, I became at once involved in a continuing series of about 20 unpiloted vehicle programs at the Flight Research Center until my retirement from NASA in 1985. The unpiloted, or remotely piloted, vehicle programs appealed especially to me

Bob Kempel, HL-10 stability and controls engineer. (NASA photo EC86 33445-1)

because they were easy to keep small and innovative and they involved conducting experiments of higher risk.

I have always felt that my talent with people is as a catalyst, a person who can help get individual team members launched creatively in different directions of exploration, especially when the venture is into new and uncharted territory. My talents at NASA seemed best used in small programs of no more than 10–15 people, the larger programs soon becoming complex matters of management and bureaucracy best left to those with talents in those areas.

NASA's HL-10 Team

Operations engineer Herb Anderson headed the 13-member HL-10 hardware team that included crew chief Charles W. Russell; mechanics Art Anderson, John W. "Bill" Lovett, and William "Bill" Mersereau; aircraft electricians Dave Garcia and Albert B. "Al" Harris; instrumentation engineer William D. Clifton; instrumentation technician Richard L. Blair; operations systems engineers Andrew "Jack" Cates and George Sitterle; and inspectors Bill Link and John Reeves. The HL-10 11-member analytical team consisted of aerodynamicist Georgene Laub; systems engineers John Edwards, Berwin Kock, and Wen Painter; stability and control engineers Robert W. "Bob" Kempel and Larry Strutz; simulation engineers Don Bacon, Larry Caw, and Lowell Greenfield; and two members of the United States Army, Lieutenants Pat Haney and Jerry Shimp.

Bob Kempel assumed the leading role in the analysis of the stability and control characteristics of the HL-10, taking over the analytical role previously performed by Ken Iliff and Larry Taylor. In developing the control laws, Kempel worked hand-in-hand with the NASA Langley wind-tunnel team and the Northrop aircraft designers. Kempel had watched the evolution of the M2-F2 configuration, and he was aware of the vehicle's marginal lateral-directional control characteristics. He swore that he would do everything he could to make the HL-10 the best flying lifting-body.

"We were the neophytes," Kempel recalled later of the tension surrounding the first flight of the HL-10. The team preparing the HL-10 simulation had only three to six years of experience. Still "untried and unproven," to use Kempel's words, the HL-10 team wasn't really a full-fledged team yet. "We were a group of individuals working as individuals toward a common goal," Kempel said. "Our approach to completing our tasks was not necessarily lacking in quality but, rather, lacking in experience."[2]

Pilots who "flew" the HL-10 real-time simulator found the vehicle's handling and lift-to-drag ratio suspiciously good, compared to those of the M2-F2. Others—including Paul Bikle, the Air Force's M2-F2 team, and NASA project manager John

2. Robert W. Kempel, Weneth D. Painter, and Milton O. Thompson, ***Developing and Flight Testing the HL-10 Lifting Body: A Precursor to the Space Shuttle*** (Washington, DC: NASA Reference Publication 1332, 1994), pp. 21–22. Since Kempel was the principal author of this paper, to avoid convoluted phraseology the narrative treats the words in it as his.

McTigue—were equally skeptical of the HL-10's simulation results. However, the simulator showed the HL-10 to be much more stable and generally much easier to handle than the M2-F2, besides having a better lift-to-drag ratio.

"We always had a difficult time convincing the pilots that we really did know what we were doing," Kempel said. "Before flight they remained skeptical. Our desire, of course, was to have simulations somewhat pessimistic rather than the other way around. We did not want to foster overconfidence."

It wasn't easy instilling even minimal confidence as "the new kids on the block," recalled Kempel. "Managers would pass us in the corridors and shake their heads." The comment most often heard was, "It can't be that good!"[3] The team's work continued, nevertheless, kept on track by Gary Layton. Despite the team's lack of assurance, all objectives were met in preparation for the first flight of the HL-10.

HL-10's Maiden Flight

Shortly before Christmas, the HL-10 team convinced Paul Bikle and the rest of NASA and Air Force management that it was ready for the first glide flight. Two captive flights of the HL-10 on the B-52 followed, allowing the team to practice going through check lists and control-room procedure, as well as correct anomalies that appeared in hardware or procedure.

On 21 December, the HL-10 was positioned beneath the B-52's right wing, lifted into position, and attached. Preflight checks were completed. However, the flight was aborted later that day due to an electrical tip-fin flap failure. Since only the subsonic configuration would be flown initially and the flaps would not be moved outboard for the first flight, the wiring was disconnected and stowed.

All preparations for the first free-flight of the HL-10 were completed early the next day, 22 December. Strapped into the cockpit, project pilot Bruce Peterson completed the preflight checks. The canopy was lowered once all ground preparations had been completed. The B-52 taxied to Edwards' main runway, Runway 4. The take-off was smooth. The flight plan called for a launch point about three miles east of the eastern shore of Rogers Dry Lake, abeam of lakebed Runway 18, almost directly over the Air Force's Rocket Propulsion Test Site (now known as the Phillips Laboratory). Launch heading was to be to the north with two left turns. The ground track looked much like a typical left-hand pattern with the launch on the downwind leg, then a base leg, a turn to final, and a final approach to landing on Runway 18.

At 10:30:50 a.m. PST, the HL-10 was launched from the B-52 at 45,000 feet and at an airspeed of 195 miles per hour. Actual launch proved to be very similar to simulator predictions. Although airplane trim was much as expected, Peterson sensed what he described as a high-frequency buffet in pitch and somewhat in roll, later

3. *Ibid.*, p. 21.

specifically identified as a "limit cycle"—that is, a rapidly increasing oscillation of a control surface that occurs when the sensitivity (or "gain") of the automatic stabilization system is too high. As speed increased, the limit cycles got noticeably worse. During the first left turn, Peterson noticed that the sensitivity of the pitch stick was excessively high. As the flight progressed, the limit cycles increased in amplitude, and it became obvious that the longitudinal stick was excessively sensitive.

Throughout the flight, Peterson and systems engineer Wen Painter were in constant communication through flight controller John Manke, making gain changes in the vehicle's stability augmentation system (SAS). During the somewhat premature landing, the SAS gains were set at the lowest rate possible without being shut off. Pitch problems masked the difficulties in the roll axis. Peterson initiated the landing flare at approximately 370 miles per hour (mph) with touchdown at about 322 mph, or about 35 mph faster than anticipated. The first flight of the HL-10 had lasted 189 seconds—that is, three minutes and nine seconds from launch to touchdown—with an

HL-10 turning to line up with lakebed Runway 18. The main part of Edwards Air Force Base is at the top of the photo and North Base is shown on the right. (NASA photo E69 21089)

average descent rate of nearly 14,000 feet per minute. Following Painter's requests for adjustments in SAS gains, Peterson had done an excellent job of flying and landing the marginally controllable HL-10.

Peterson remained greatly concerned about the pitch sensitivity and limit cycles. To be precise, a limit cycle is a condition in a feedback control system that produces the uncontrollable oscillation of a control surface due to closed-loop phase lag that, in turn, results from excessive lag in the system (called "hysteresis"), accumulated free play of mechanical linkages, and power actuator non-linearity. The amplitude of the cycle increases with each augmentation to airspeed and system gain setting.

The particular limit cycle that occurred during the first flight of the HL-10 was a 2.75 Hz oscillation (0.4g peak-to-peak) feeding through the gyro-driven SAS. Primarily the problem was in the pitch axis, although it also affected the roll axis. The problem was more severe during the final third of the flight, despite the fact that the SAS gain had been reduced from 0.6 to 0.2 deg/deg/sec. Afterwards, for the entire first HL-10 flight, Peterson gave the pitch axis a Cooper-Harper pilot rating of 4, a rating indicating that deficiencies warrant improvement and are not satisfactory without improvement.

The flight proved to be a large disappointment for the HL-10 team. It seemed to confirm the opinion of others who had said that the team didn't know what it was doing. The team's morale was at low ebb, the flight results quite poor in comparison with the expected results of preflight simulation and analysis.

After the holidays, as 1967 began, team members concluded that if they fixed the stick sensitivity and lowered the SAS gains, they could probably try another flight. There was, however, one lone dissenter in the group. Systems engineer Wen Painter was not convinced that the team completely understood all of the problems.

Continuing to analyze the results of the first flight, Painter argued against another attempted flight, despite the fact that Bruce Peterson had convinced Bikle that the team should try again. Their confidence shaken by the first flight's results, the team gave in to Painter. Bikle backed Painter fully, saying that if Painter didn't sign the ship's book—that is, okay the flight—there would be no flight. Following Painter's suggestion, the team initiated an in-depth unified analysis of the data from the first flight. Very subtly this effort would mold them over time into a real team of proven experience.

Post-Flight Analysis

Two serious problems identified even before touchdown were substantiated in post-flight analysis: large amplitude limit cycles in the pitch SAS and extreme sensitivity in the longitudinal stick.

The problem with limit cycles apparently was caused by higher-than-predicted elevon control effectiveness and feedback of a 2.75 Hz limit-cycle oscillation through the SAS. The solution involved using lower SAS gains and modifying the structural resonance 22 Hz mode lead-lag filter that had been installed before the first flight. The modification consisted of a lead-lag network in the SAS electronics and a notch filter, a device that removes a nuisance frequency while having relatively little effect on lower and higher frequencies.

The problem with longitudinal stick sensitivity was relatively simple to solve with a basic gearing modification. On the first flight, the stick gearing of 6.9 deg/inch of elevon proved to be much too sensitive. The nonlinear gearing used in flights 10-37 was approximately 3.5 deg/inch in the elevon range for landing—or about half of what it had been during the first flight. This type of problem is easy to miss when all preparations for flight are made on a fixed-base engineering simulator, a "safe" environment that is relatively relaxed for the pilots who know that if anything goes wrong, they can simply reset the computers. Furthermore, the trim characteristics of a new aircraft are not known precisely. Stick sensitivity, whether longitudinal or lateral, has always been difficult to determine in fixed-base simulations. Pilots always want a very responsive aircraft.

A third problem proved more elusive, not apparent to the pilot or test team during the initial post-flight analysis: lack of longitudinal or lateral-directional control at some portions of the flight. Peterson had realized during the first flight that something wasn't right at high gains and consequently had flown a faster landing approach. Understanding and resolving this problem would require more thorough flight investigation and the assistance of NASA Langley, grounding the HL-10 for fifteen months.

In-Depth Flight Investigation

Wen Painter had insisted that even more analysis needed to be done to find out why lateral control was good sometimes and almost totally lacking at other times, so Bob Kempel launched an in-depth investigation. The assumption before the first flight of the HL-10, according to Kempel, had been that the simulation generated from wind-tunnel test results, an analog computerized mathematical model of the HL-10, was relatively accurate in representing the actual flight vehicle. The expectation, then, was that if flight-recorded control inputs were fed into the computerized model, the dynamics (or motions) of the simulator should be similar to those of the actual vehicle—a technique used for years to validate aerodynamic data by actual flight data. Ideally, the simulation matches the flight exactly; however, such perfection is rarely realized. When the simulation and flight data don't match, aerodynamic parameters are adjusted to duplicate as closely as possible the flight motions. In this way, engineers can then determine how wind-tunnel aerodynamics differ from flight and, perhaps, even why they differ.

The first engineering task in the in-depth flight investigation involved selecting twelve specific maneuvers from five to fifteen seconds in duration from the flight

Wen Painter, HL-10 systems engineer who promoted the engineering investigation resulting in the grounding (and aerodynamic fix) of the HL-10 for a 15-month period after its first flight. (NASA photo EC79 11441) p. 71

results. Next, the engineers tried to match these maneuvers with those generated by computer, a good match being one in which the computer solution overlays all parameters recorded during flight within the specified time interval and there is little difference between the flight maneuver and the computer generation. However, there were no good matches and only seven found to be acceptable. The other five maneuvers were impossible to match by model. Kempel and the team determined that the computer solutions didn't even remotely resemble the actual flight response of the HL-10. They concluded that they must not have been using an accurate mathematical model, leading them to examine once more the actual flight data.

We decided to play the entire flight-recorded data back through the ground station, the team this time selecting parameters that would be grouped together. The team selected three families of specific data—accelerations, angular rates next to the control inputs, and information from control surface strain gauges. We then traced out these groupings as a function of time. The new approach gave the team the capability of looking at eight channels of data on each strip-chart. What we found was quite revealing.

The inexperience of the team had shown in how it had earlier arranged the control-room strip-charts for the initial post-flight analysis. Real-time data hadn't been arranged in the best logical manner for accurate assessment of data families. With the data re-arranged, the team found that, although of different parameters, each of the traces generally moved with the appropriate responses indicating the vehicle's motion. However, during certain portions of the flight, some of the traces would become blurry or fuzzy, especially the control surface strain gauges when a higher frequency disturbance occurred. When the data was lined up on a common time interval, many data traces displayed similar phenomena.

A second but related discovery was that there had been two significant intervals when Bruce Peterson had commanded significant amounts of aileron, only to have the vehicle not respond until the angle of attack was reduced. Peterson was disturbed enough by the vehicle's response to control input that Kempel and the team decided to investigate it further. What they found was that each time the problem occurred, the angle of attack was above the range of 11 to 13 degrees, and that as the angle of attack decreased through this range, the ailerons suddenly became very effective, producing significant amounts (30 to 45 degrees per second) of roll angular rate.

When the team computer-matched these two time intervals, the initial part of each response would not match. However, as the angle of attack was reduced to the point that the ailerons became effective, the mathematical model began to match the flight data. But why?

As Kempel recalls, "We began to think that a massive flow separation was possible over the upper aft portion of the vehicle at the higher angles of attack, causing the control surfaces to lose a large percentage of their effectiveness. . . . This flow separation can be likened to the sudden loss of lift and increase in drag of a conventional wing as AOA [angle of attack] is increased and the wing stalls. As the AOA was decreased, the airflow would suddenly reattach and the controls would behave in their normal fashion. The more we looked at the data, the more plausible this theory seemed; although the wind-tunnel data did not indicate a problem to the degree that we had experienced in flight. The data also indicated a significant loss of lift-to-drag ratio above Mach numbers of 0.5 and AOA of 12 degrees. This finding further convinced us that the problem was caused by massive flow separation."

At this point, Kempel and his team decided to share their preliminary findings with the NASA Langley engineers since, as Kempel said, the HL-10 was "their 'baby.'" The Langley team agreed to do more wind-tunnel tests immediately, using the 0.063-scale, 16-inch-long HL-10 model. According to Kempel, the Langley team's decision seemed "highly unusual because, typically, wind-tunnel schedules are made at least a year and, sometimes, [several] years in advance."[4] As the Langley team urged them to do, Kempel and his team packed their data and bags and traveled to NASA Langley to work jointly on the situation.

Bob Kempel, Berwin Kock, Gary Layton, and Wen Painter of the Flight Research Center gathered around a table with Langley's Bill Kemp, Linwood (Wayne) McKinny, Bob Taylor, and Tommy Toll in the building housing Langley's 7-by-10-foot high-speed wind tunnel. Kempel and his team, after presenting their data, theorized that the problem was caused by massive flow separation. Bob Taylor jumped up from his chair, angrily slammed his mechanical pencil to the floor, and let loose with a string of oaths. After he calmed down, Taylor said that he had earlier thought that this would be a problem. He had had a gut feeling that the flow separation seen by the Langley team on the wind-tunnel model would be worse in flight, and he was upset with himself for not following his instincts as an aerodynamicist and adding preventative measures to the HL-10 design before the vehicle was built.

The discussion then turned to what could be done now. The Langley team agreed to give the problem its immediate attention, assuming responsibility for coming up with a remedy. Kempel and his team left Langley more aware than they had been earlier of why they were having a lateral control problem in flying the HL-10. They agreed that, until Langley came up with a solution, the HL-10 would not be flown. While they waited for word from Langley, they busied themselves with solving the problems they had determined earlier (stick sensitivity and limit cycles), enlisting the help of Northrop in designing the electronic notch filter for eliminating the limit-cycle mode from feeding back through the flight control system.

HL-10 as "Hangar Queen"

The HL-10 was a "hangar queen" for the next 15 months, grounded after its first flight three days before Christmas 1966. During this time, flight safety began receiving more attention, to some extent due to the near crashes and temporary losses of control with the other lifting bodies. Adherence to flight schedules took a second priority to flight safety, benefiting the HL-10 program. Bob Kempel was given free license to work without a time restraint in leading the effort to fix the vehicle's control problems.

Throughout the winter and spring of 1967, members of the NASA Langley team continued to work on correcting the flow-separation problem, coordinating their efforts with those of Kempel and his team at the Flight Research Center. The Langley team came up with two possible ways to fix the problem, both modifications concentrating

4. *Ibid.*, p. 26, for quotations 2 and 3.

on changes to the outboard vertical fins. The first proposed modification involved thickening and cambering the inside of the fins. The second proposed slightly extending and cambering the leading edges. Langley ran a full set of wind-tunnel tests on both proposed modifications, sending the resulting data to Kempel and his team. Although the Langley team members gave their assessment of the wind-tunnel results, they left the decision of which modification to use up to Kempel and the team at the Flight Research Center.

Kempel recalls that once he had the preliminary data from these wind-tunnel tests, he initiated his own extensive evaluation of the data. "Preliminary data," Kempel clarifies, "was the wind-tunnel guys' way of telling us that they had worked most of their magic in data reduction, but that they still were not going to say that this was the last word."[5] Kempel plotted all of the data from digital listings by hand. Although engineers today use computer plotting routines to do what Kempel in 1967 had to do by hand, the approach made him and other team members intensely familiar with the data, for the extensive process of hand-plotting meant they had to live with the data day in and day out.

During the summer of 1967, Kempel plotted all of the data for both proposed modifications as a function of angle of attack for constant Mach numbers. He made all plot scales uniform to ease comparisons, plotting thousands of points in this way. Once the data was lined up and compared, Kempel found there were some subtle but significant differences between the Langley wind-tunnel data and the data set generated by the HL-10 simulator at the Flight Research Center.

As Kempel explains it, "Some non-linearities in the original data were not present" in the Langley data. He hypothesized that "if these non-linearities indicated flow separation, then the lack of these would indicate no flow separation or separation to a lesser degree."[6] Based on that theory, Kempel backed using the second modification proposed by Langley. He presented his hypothesis to his boss, aerodynamicist Hal Walker, and then to the management at the Flight Research Center. With their agreement and the concurrence of the NASA Langley team, Kempel and his team began making arrangements for the modification of the HL-10.

In the early autumn of 1967, Northrop Norair was contracted to design and install the modification that would be the final configuration change to the HL-10. Northrop and NASA decided that the modification would involve a fiberglass glove, backed by a metal structure. Work on the glove continued through the autumn and winter of 1967.

As Kempel recalls, "In the NASA hangar, Northrop's Fred Erb shed his normal working attire—a suit—and donned coveralls to assist in the installation of the fiberglass glove. He was a senior-level engineer with over 25 years with Northrop, rolling

5. *Ibid.*, p. 27.
6. *Ibid.*, pp. 27–28.

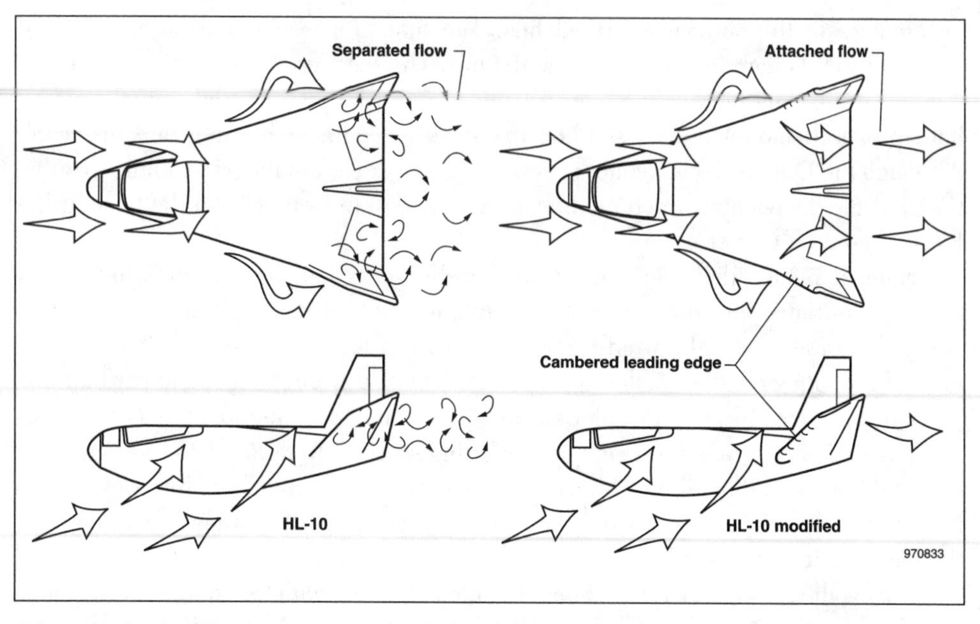

Schematic showing HL-10 aerodynamic modification (original drawing by Dale Reed, digital version by Dryden Graphics Office).

up his sleeves and getting his hands dirty."[7] "A real engineer!", Kempel might have added.

By the spring of 1968, the HL-10 was nearly ready to end its stay as a hangar queen, with vehicle preparation then in its final stages. Changes in the configuration, flight controls, and internal systems were already finished.

Gentry, Peterson, and the M2-F2

Meanwhile, following Jerry Gentry's flight on 14 November 1966, the M2-F2 was grounded five and a half months so that the LR-11 rocket-propulsion system could be installed by the lifting body's team under the leadership of Meryl DeGeer. Gentry made four glide tests in the M2-F2 by 2 May 1967, conducting research maneuvers to define the vehicle's aerodynamic characteristics and preparing for planned rocket-powered supersonic flights. Having flown the M2-F2 successfully several times, Gentry was by this time firmly established as an experienced lifting-body pilot, soon becoming the Air Force's most active pilot in the joint NASA-Air Force lifting-body program.

7. *Ibid.*, p. 28.

A key member of the M2-F2 team, Gentry knew each crew member personally. Practical jokes abounded between them, and Gentry never once let anyone forget that he represented the Air Force on the project. During his early flights in 1966, he had told the crew that he hated the zinc-chromate yellow-green color of the insides of the lifting bodies. Afterwards, during one of his flights, his flight-line car, a 1954 Ford, was "borrowed" long enough to be painted entirely in zinc-chromate yellow-green at the NASA paint shop.

In retaliation, Gentry and his Air Force cronies sliped over to NASA during one early morning to paste a large Air Force sign on the side of the HL-10, which originally had no markings indicating Air Force involvement in the program. When the NASA crew members arrived and saw the Air Force sign, they promptly removed it. Later, they had the last word, decorating Gentry's yellow-green Ford by pasting large "flower power" decals all over it, the decals then popularly in use mainly by the era's "flower children."

By the winter and spring of 1966–1967, the two official lifting-body pilots—the Air Force's Jerry Gentry and NASA's Bruce Peterson—were doing alternate flights in the lifting bodies. Since Peterson had flown the HL-10 for its maiden flight on 22 December 1966, it was Gentry's turn to fly the M2-F2 on 2 May 1967 for its first flight with the rocket system installed. On this glide flight, his fifth in the M2-F2, Gentry reported that the weight increase from the installed rocket system had not changed the vehicle's control characteristics. However, he also confirmed what Milt Thompson and Bruce Peterson had reported on their previous flights: that if the M2-F2 is not flown properly, loss of roll control can occur quickly.

In September of 1966, during the symposium of the Society of Experimental Test Pilots, Bruce Peterson had given a detailed description of the M2-F2's lateral control characteristics. Maneuverability "was not appreciably affected" as yaw and roll damper gains were reduced to zero during the first 180 degrees of approach on the fifth flight, he said. However, he felt at the time that "abrupt aileron or rudder inputs could readily induce Dutch roll oscillations"; and these "could be continuous and could seriously hamper the pilot in holding a bank angle." His strategy was to "nudge" the M2-F2 to the desired bank angle by using small lateral control inputs.

"Acceptable lateral control is achieved only by means of aileron-rudder interconnect since the adverse yaw due to aileron at most flight conditions results in roll reversal," he said.

> "The optimum interconnect ratio varies with angle of attack . . . and dynamic pressure. Consequently, unless the pilot is willing to change interconnect continuously throughout the flight, roll effectiveness varies from sluggish to extremely sensitive, bordering on a pilot-induced oscillation. Even at the optimum interconnect ratio, the response to lateral control input is not smooth regardless of magnitude or rate of input. This is due to the initial rolling moment produced by the rudder through the interconnect, which is the opposite of the desired roll direction. Vehicle response to lateral control input is always somewhat of a surprise to the pilot in terms of lag and resultant initial rate."[8]

Crash of the M2-F2

On 10 May 1967, eight days after Gentry's glide flight, it was Bruce Peterson's turn for a glide flight in the M2-F2 with the rocket system installed. It had been eight months since Peterson's last six-minute glide flight in the lifting body, and this would be his third M2-F2 flight.

All went well during the beginning of Peterson's flight on 10 May. He launched away from the B-52 at 44,000 feet, heading to the north, flying east of Rogers Dry Lake, and descended at a steep angle to 7,000 feet. Then, as he flew with a very low angle of attack, the M2-F2 began a Dutch roll motion, rolling from side to side at over 200 degrees per second. Peterson increased the angle of attack by raising the nose. The oscillations stopped, but now the M2-F2 was pointed away from its intended flight path. Realizing that he was too low to reach the planned landing site on lakebed Runway 18, Peterson was rapidly sinking toward a section of the lakebed that lacked the visual runway reference markings needed to accurately estimate height above the lakebed.

At this moment, a rescue helicopter suddenly appeared in front of the M2-F2, distracting Peterson who was still stunned and disoriented from the earlier Dutch roll motions. He radioed, "Get that chopper out of the way." A few seconds later, he radioed, "That chopper's going to get me." NASA pilot John Manke, flying

8. Quotations in two paragraphs above from Bruce Peterson's comments in Milton O. Thompson, Bruce A. Peterson, and Jerauld R. Gentry, "Lifting Body Flight Test Program," *Society of Experimental Test Pilots, Technical Review* (September 1966): 4–5.

M2-F2 after the crash, showing how the cockpit was held above the ground by the rollover structure. Bruce Peterson's helmet is in the foreground. (NASA photo E67 16731)

chase in an F-5D, assured Peterson that he was now clear of the helicopter, which had chugged off out of Peterson's flight path.[9]

Trying to buy time to complete the flare, Peterson fired the landing rockets. The M2-F2 flared nicely. He lowered the landing gear, only one-and-a-half seconds being needed for the M2-F2's gear to go from up and locked to down and locked. But time had run out. The sudden appearance of the helicopter likely had distracted Peterson enough that he began lowering the landing gear half a second too late.

Before the gear locked, while it was still half-deployed, the M2-F2 hit the lakebed. The weight of the vehicle pushed against the pneumatic actuators, and the landing gear was pushed back up into the vehicle. The round shape of the vehicle's bottom did not lend itself to landing minus landing gear. The result was more like a log rolling than a slide-out on a flat bottom. (By contrast, the shape of the HL-10 likely would have lent itself readily to a gear-up landing, had one been required. Langley engineers had even given serious thought to eliminating the HL-10's landing gear for spacecraft recovery.)

9. Quotations from Hallion, *On the Frontier*, p. 159; Wilkinson, "Legacy of the Lifting Body," pp. 57-60, but Dale Reed was watching the whole episode on a TV monitor from the control room, as pointed out below in the narrative, so he heard the comments first hand.

As the M2-F2 contacted the ground, the vehicle's telemetry antennae were sheared off. As this happened, I and the other engineers in the control room watched the needles on instrumentation meters flick to null. Startled, we looked up at the video monitor in time to see the M2-F2, as if in a horrible nightmare, flipping end over end on the lakebed at over 250 miles per hour. It flipped six times, bouncing 80 feet in the air, before coming to rest on its flat back, minus its canopy, main gear, and right vertical fin. The M2-F2 sustained so much damage that one would have been hard pressed to identify it visually as the same vehicle.

By all odds, Peterson could have been expected to have died in the crash. He was seriously injured. Assistant crew chief Jay King quickly crawled under the M2-F2 to shut off the hydraulic and electrical system. He found Peterson trying to remove his helmet. King unstrapped him and helped him out of the vehicle. Peterson was rushed to the base hospital at Edwards for emergency care. Afterwards, he was transferred first to the hospital at March Air Force Base near Riverside, California, and later, to UCLA's University Hospital in Los Angeles.

The heavy metal cage-like structure around the cockpit—ironically, added to the M2-F2 by its NASA/Northrop designers simply to provide ballast and save their pride—was mainly what saved Peterson's life. Even with this added protection, his oxygen mask was ripped off as his head made contact with the lakebed. Each time the vehicle rolled, a stream of high-velocity lakebed clay hammered at Peterson's face. He suffered a fractured skull, severe facial injuries, a broken hand, and serious damage to his right eye. He underwent restorative surgery on his face during the ensuing months; however, he later lost the vision in the injured eye from a staphylococcus infection.

He returned to the NASA Flight Research Center as a project engineer on the CV-990, F-8 Digital Fly-By-Wire, and F-8 Supercritical Wing. He continued to fly in a limited way on the CV-990 and F-111 and eventually became the Director of Safety, Reliability, and Quality Assurance. He also continued to fly as a Marine reservist. Later, he left NASA to serve as a safety officer at Northrop in the flight tests of the B-2 bomber and other aircraft.

About two years after the crash of the M2-F2, the popular television series ***The Six-Million-Dollar Man*** began its six years of weekly programming, using NASA ground-video footage of the crash as a lead-in to each episode. The producers of the television series capitalized on Peterson's misfortune by inventing a "bionic man" (played by Lee Majors) who had missing body parts replaced with bionic devices. Colonel Steve Austin, the fictional television character played by Majors, had, like Peterson, also lost an eye in the crash.

As can happen only in Hollywood, the fictional Austin gained a bionic eye with super powers. The television show also multiplied the injuries of Austin beyond those suffered in real life by Peterson, giving him two bionic legs and a bionic arm that provided him with super power and speed. Nevertheless, NASA pilot Bruce Peterson is the real-life model on which ***The Six-Million-Dollar Man*** is based. Due to the popularity of this television series, it's possible that as many Americans viewed the crash of the M2-F2 on television as later viewed the first televised NASA shuttle landings.

The crash of the M2-F2 was the only serious accident that occurred during the twelve-and-a-half years of flight-testing eight different lifting bodies.[10] Because of the popularity of the television program ***The Six-Million-Dollar Man***, most people are more familiar with the solitary serious accident that occurred during the lifting-body program than they are with its extensive record of otherwise accident-free success.

10. These were the M2-F1, M2-F2, M2-F3, HL-10, HL-10 modified, X-24A, X-24B, and the Hyper III.

CHAPTER 6

BACK TO THE DRAWING BOARD

The crash of the M2-F2 left us with no lifting bodies to fly for almost a year. When the M2-F2 crashed in early May 1967, the HL-10 had been a hangar queen for over four months, and it would remain grounded for another eleven months while its aerodynamic problems were fixed before its second flight. Bikle had grounded the M2-F1 permanently, the "flying bathtub" that had launched the lifting-body effort four years earlier now destined to be a museum artifact. Another lifting body was in the works, the Air Force Flight Dynamic Laboratory at Wright-Patterson Air Force Base having a contract with the Martin Aircraft Company of Middle River, Maryland, for designing and building a piloted lifting body originally designated the SV-5P and later known as the X-24A. However, it would be another two years before it was ready to fly.

Despite the setbacks in lifting-body flight testing, competition continued to flourish between the flight-test teams of the NASA/Air Force M2-F2 and the NASA HL-10. With the Air Force and three different NASA sites—Ames on the M2-F2 and Langley on the HL-10, each in conjunction with the Flight Research Center—actively involved on the M2-F2 and in flight operations for the HL-10, the dynamic energy of their interaction could have been destroyed within the multiple organizational channels through which it had to travel. It was amazing to watch these teams cut across NASA and Air Force channels and remain unified, their first allegiance being to their shared lifting-body project.

Rebirth of the M2-F2

The crashed M2-F2 was pathetic-looking, nearly no skin panels without dents or damage. Rather than scrapping the M2-F2, John McTigue had the vehicle sent to Northrop's plant in Hawthorne, California, where Northrop technicians put the battered vehicle in a jig to check alignment, having removed the external skin and portions of the secondary structure, and then removed and tested all systems and parts, an inspection process that took the next two months. Many parts such as valves and tanks were tested at the Flight Research Center's rocket shop. Meanwhile, the M2-F2 team tackled the difficult problem of fixing the vehicle's control problems. Over the next 60 days, the NASA Ames team members, led by Jack Bronson, gave high priority to wind-tunnel tests for finding that solution. Using a make-shift model of the M2-F2, they tried five different approaches to fixing the problem with elevon adverse yaw.

First, they tried canting the elevon hinge lines so that side force directly on the elevons would give favorable yaw into a turn. This approach failed, because there were still more pressure effects on the vertical fins that offset any favorable pressure on the elevons.

Second, they tried an extra horizontal surface with two elevons attached between the right and left vertical tail tips, putting favorable pressures on the vertical tails that would reverse the yawing moments. This approach was abandoned due to its complexity and its structural problems.

Third, they tried converting the elevons to a bi-plane arrangement with standoffs supporting a second horizontal surface above each elevon so that the original elevons and standoff surface would move as a control unit. This approach was abandoned because it did not produce the favorable pressure gradients they had hoped it would.

Fourth, they tried extending the elevons aft of the body, away from the vertical fins. This approach succeeded in eliminating about half of the adverse yawing moments, although it also became apparent that pressure gradients were being affected upstream near the vertical tails from elevon deflection.

Finally, they tried installing a center fin that would act as a splitter-plate between the right and left elevons, producing side forces that would counter those of the outer vertical fins. For example, following a right roll command by the pilot, the original M2-F2's right elevon trailing edge moved upward. The pressure field on the upper right side of the body would increase due to this deflection, pushing down on the right side of the body. This increased pressure would also push on the inner side of the right vertical tail, pushing the tail to the right and the nose to the left, resulting in adverse yaw. With the center fin installed on the M2-F2, however, this pressure would also push against the right side of the center fin, opposing the adverse yaw effects from the pressure pushing to the right against the right vertical tail and, as a result, canceling the moments of adverse yaw.

Jack Bronson's team at NASA Ames ran wind-tunnel tests on center fins of various sizes. As expected, the larger ones produced more proverse (favorable) yaw than did smaller ones. Meryl DeGeer, the M2-F2 operations engineer at the Flight Research Center, was asked to provide a clearance drawing of the largest vertical fin that would fit under the B-52 pylon. As it turned out, the M2-F2/B-52 adapter could not be used if a center fin were installed on the M2-F2, for it had a large beam running down the center. However, DeGeer and the Northrop designers decided that the HL-10 adapter—with a slight modification—could be used for both vehicles since it had been built to accommodate the center fin on the HL-10. NASA Ames tested the fin shown in DeGeer's drawing, and it worked. The fin not only neutralized the adverse yaw effects but it also produced a small amount of proverse yaw beyond what was needed to cancel adverse yaw.

A conference called by Gary Layton was held at the NASA Flight Research Center, attended by team members from both NASA Ames and the Flight Research Center as well as the Air Force. Due to the wind-tunnel test results, the center fin was unanimously accepted by the attending team members as the way to fix the control problems on the M2-F2. The NASA Ames team then gathered a more complete set of data on the new configuration. The team at the Flight Research Center analyzed the Ames data that showed the elevons to have a small amount of proverse yaw, modified the M2-F2 simulator, and calculated new root-locus characteristics.

Bob Kempel remembers making some root-locus calculations on the old and the new M2-F2 configurations at that time. He found the difference in controllability to be as extreme as that between night and day. The new configuration with the center fin had good roll control characteristics with no tendencies for problems in pilot-induced oscillation (PIO). Although Kempel was officially on the HL-10 team at the time, he had a vested interest in the M2-F2 from having done some analysis on it early in its development. Never happy with the lateral control-system design on the original M2-F2, he had aligned himself with the HL-10, which he originally considered the better of the two heavyweight lifting bodies. With the center fin added to the M2-F2, Kempel agreed that the vehicle could become a good flying machine.

As their main mathematical tools in analyzing all motions made by an aircraft during flight, stability and control engineers such as Bob Kempel use La Place transforms, differential equations, and linear algebra. Winged aircraft normally have such typical motions as roll, spiral, and Dutch roll modes. Lifting bodies, on the other hand, can have a unique motion called a coupled roll-spiral mode, which Kempel documented on the M2-F2 in September 1971 in a NASA report entitled, "Analysis of a Coupled Roll-Spiral-Mode, Pilot-Induced Oscillation Experienced With the M2-F2 Lifting Body."[1] Kempel explains that the oscillatory coupled roll-spiral mode results from a combination of non-oscillatory roll and spiral modes. When poor roll controls such as the M2-F2 elevons are used, PIO problems result.

The control problems in piloting a lifting body are somewhat like the control problems experienced by a lumberjack in maintaining his balance during the sport of log-rolling, something I know a little bit about from growing up near the logging industry in Idaho. A log is similar to a lifting body in that both are very slippery in a roll, neither having anything like wings that work to resist the rolling motion in water, for the log, or in air currents, for the lifting body. A lumberjack wearing spiked boots has a pair of good controls on the log he's rolling. With constant attention, he can use his spiked boots to control the log's motion. Were the lumberjack wearing instead a pair of ordinary slick-soled shoes, however, he'd have only a pair of poor controls to use. Even with constant attention, he'll eventually lose control of the log he's rolling and, when a wave (analogous to a side gust on a lifting body with poor controls) hits the log, he's going to get very wet.

By 1967, we had flown two lifting-body configurations and were about to fly a third, the M2-F3, the rebuilt M2-F2 with the added center fin. The log-roller analogy

1. R. W. Kempel, "Analysis of a Coupled Roll-Spiral-Mode, Pilot-Induced Oscillation Experienced With the M2-F2 Lifting Body" (Washington, DC: NASA Technical Note D-6496, 1971).

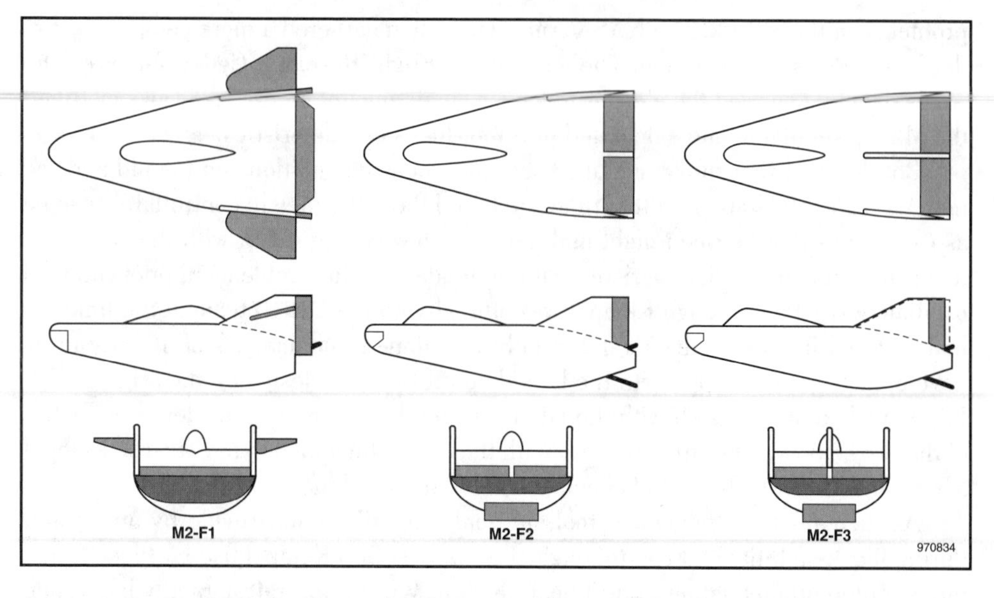

Schematic showing the M2-F1, M2-F2 and the M2-F3 (original drawing by Dale Reed, digital version by Dryden Graphics Office).

applies as well to the differences among the M2-F1, the M2-F2, and the M2-F3. The M2-F1 had the large "elephant ears", the external elevons, that provided good roll control, similar to the lumberjack wearing the spiked boots. The "elephant ears" also served as flat surfaces that slow down, or damp, rolling motions, similar to what would happen if the lumberjack nailed a board to the log. When we went from the M2-F1 to the M2-F2 configuration, we essentially deprived the lumberjack of his spiked boots and removed the board from the log, depriving him of the means for good roll control and damping. When we converted the M2-F2 to the M2-F3 configuration with the center fin, we essentially gave back to the lumberjack his pair of spiked boots, equipping him with the means for good roll control. However, our lumberjack would still have a slick log with no way to slow down (damp) the rolling motions minus the board nailed to the log. What the board nailed to the log provides the lumberjack, a stability augmentation system (SAS) on a lifting body provides the pilot, both helping to damp oscillations and other quick movements.

Birth of the M2-F3

Northrop was enthusiastic about wanting to rebuild the M2-F2 into the M2-F3, strengthening the resolve of the NASA teams to seek approval from NASA Headquarters for continuing the M2 program. NASA Headquarters was reluctant about authorizing more M2 flight tests, but project manager John McTigue was not one to give up easily. Eventually his tenacity succeeded in getting NASA's Office of Advanced Research and Technology to authorize Northrop in March 1968 to continue its "inspection" of the wrecked lifting body. The Northrop team that had built the M2-F2 was still intact, soon to be transferred onto other Northrop projects, so this was the last opportunity we had to have the vehicle rebuilt at low-cost, using the best possible Northrop team for the job.

The lifting-body program was also fortunate to have the help of Fred DeMerritte to keep the effort going at NASA Headquarters. DeMerritte and McTigue had an unwritten agreement that they would proceed quietly at a steady pace until the M2-F3 was ready to fly. McTigue had Bikle to back him up at the Flight Research Center, but DeMerritte was on his own at NASA Headquarters. There was no official authorization for conducting an M2-F3 flight program; however, DeMerritte managed to find a way to continue sending money in incremental amounts to John McTigue to keep the "inspection" going until official approval was obtained.

Just how tense the situation was around DeMerritte at NASA Headquarters in regards to the M2-F2/M2-F3 project is suggested by a conversation that Meryl DeGeer recalls having with DeMerritte on a visit the latter made to the Flight Research Center. DeMerritte privately asked DeGeer how things were going on the project. DeGeer said that everything was going fine but that if DeMerritte would give them some more money, they could have the M2-F3 ready to fly all the sooner. DeMerritte asked DeGeer not to push him, for then he'd be forced to say no to the project. It was nearly ten months later—on 28 January 1969—that NASA Headquarters officially announced that the Agency would repair and modify the M2-F2, returning the vehicle to service as the M2-F3, a process that took three years and cost nearly $700,000.

Since there wasn't enough money to contract out all of the work, most of the installation of systems was done by the "Skunk Works" at the Flight Research Center, similar to what had been done with the M2-F1. However, McTigue was able to retain from Northrop five engineers and nearly a dozen technicians to work with the Flight Research Center in fabricating the M2-F3 from the remains of the M2-F2.

Northrop's Fred Erb coordinated the Northrop technical effort while Meryl DeGeer, as NASA's M2-F3 operations engineer, headed up the rebuilding project at the Flight Research Center. Special design problems and parts that had to be manufactured at the Northrop facility were handled through Erb. To keep costs down, as much of the rebuilding as possible was done in the FRC shops. Working from Northrop drawings, LaVern Kelly and Jerry Reedy built new vertical tails for the M2-F3 in the FRC sheet-metal shop, two sheet-metal workers from Northrop at times assisting the shop technicians. The FRC machine shop remanufactured broken parts, including the landing gear. Rocket, fuel-system, and plumbing parts were built in the Center's rocket shop. The FRC aircraft electrical shop put together and installed the vehicle's wiring bundles and electrical systems. Besides the new central fin, a number of internal improvements and other additions were made to the M2-F3. For example, heavy components were moved farther forward, avoiding the need for nose ballast, and small changes in the cockpit area improved visibility and access to the controls. For a cleaner installation, we also rotated the LR-11 rocket engine 90-degrees.

NASA hoped that the new hydrogen-peroxide jet-reaction roll-control system installed on the M2-F3 might be used as well on future lifting-body spacecraft so the pilot could rely on a single control system from orbit to landing, rather than the multiplicity of systems used on such aircraft as the X-15. NASA planned to use the M2-F3 as a testbed for research on the lateral control problems of lifting bodies. If we could eliminate the elevons and rudders, replacing them with reaction rocket controls, we would need only one flap on the bottom of the vehicle for longitudinal trim.

According to Air Force pilot Jerry Gentry, the transformation of the M2-F2 into the M2-F3 changed "something I really did not enjoy flying at all into something that was quite pleasant to fly."[2]

HL-10 Returns to Flight

Meanwhile, after fifteen and a half months of wind-tunnel tests, simulation, control-system analysis, and modification of the outer tail fins, the HL-10 was returned to flight. Jerry Gentry flew the HL-10 for the second time on 15 March 1968, launched from 45,000 feet at Mach 0.65. From B-52 launch to touchdown, total flight time was approximately 4.4 minutes.

"I think the whole Center came out to watch this flight," recalls Joe Wilson. "People were standing on the roof, by the planes [on the ramp, and at the edge of the] lakebed. I haven't seen so many observers for a first flight since I've been here. The day was almost absolutely clear and you could see the contrails of the B-52 and [the] chase [planes] . . . two F-104s, one T-38 and the F5D. On [the] drop, everything was O.K., and for a short time you could follow the contrails. The contrails began to pop in and out [of sight], and then were gone from view."[3]

The flight plan called for mild pitch and roll maneuvers to 15-degrees angle of attack to evaluate the possibility of control degradation of the sort experienced during the first flight. To assess potential flare characteristics, Gentry executed a simulated landing flare to 2G at altitude.

A camera had been installed on the tip of the vertical fin to provide in-flight photographs of the right inboard tip-fin flap and right elevon. These surfaces had been "tufted" so that a qualitative assessment of the aerodynamic flow field could be made from the photographs. "Tufting" involves taping the ends of short pieces of wool yarn, called "tufts," on suspected problem surfaces of an aircraft for assessing the quality of airflow. If the flow is attached, the tufts lie flat in the direction of the flow across the surface. If the flow is separated, the tufts dance and flutter randomly. Generally, the conclusions following the flight were that the airflow did not separate significantly and

2. The quotation also appears in Wilkinson, "Legacy of the Lifting Body," p. 61.
3. Personal diary of NASA Flight Research Center employee Ronald "Joe" Wilson, entry for March 15, 1968, copy available in the Dryden Flight Research Center History Office.

consequently that there had been no degradation of control. (When the airflow over control surfaces separates significantly, the control is degraded because it operates aerodynamically.) However, some over-sensitivity in pitch control was observed.

In the debriefing room following the flight, Gentry said the vehicle felt solid. It had no problems in roll sensitivity. It had good longitudinal stability. He also said that, on turning to final approach, flare, and landing, the HL-10 was better than the F-104. He reported that he had put the gear down somewhere after 250 knots and had felt a sharp jolt as the nose gear touched down.

Bob Kempel, Wen Painter, and the rest of the team were as proud as peacocks following the second flight of the HL-10. When someone asked him what kind of problems had occurred on the flight, Kempel said there had been no problems at all, that the flight was a complete success from everyone's point of view. The sensitivity of the longitudinal stick, noted during the flight, was considered acceptable.

The dynamics of the HL-10 in flight proved to be as good as had been indicated by the simulator. After the second flight, Kempel said, the HL-10 attracted the attention of the pilots. "From this point on, all the pilots wanted their shot at flying the HL-10."[4]

After pilots establish confidence in a new aircraft and have a little more time to evaluate things, they often change their opinions. The situation was no different with the HL-10. Although no major modifications were required, minor adjustments continued to be made to the HL-10 throughout the remainder of the program. The HL-10 had 35 more successful flights, piloted by NASA's Bill Dana and John Manke and the Air Force's Jerry Gentry and Pete Hoag.

F-104 Used in Pilot Training

During 1968, pilots were becoming very dependent on the ground-based simulator for developing flight procedures and becoming as familiar as possible with the flight characteristics of the lifting bodies. Actual flight experience in the lifting bodies could not be relied upon to provide adequate pilot training because the typical flights were short—five to six minutes for glides, 10 to 15 minutes for rocket flights—and weeks or even months separated flights. Furthermore, for the lifting-body pilots, the first launch off the B-52 hooks was like being thrown into deep water for the first time: you either swim or sink.

In 1957–58, a young research pilot at the Flight Research Center by the name of Neil Armstrong—who, as a NASA astronaut, would later become the first human being to walk on the moon—had conducted a series of flights tests on the NASA F-104 designed to simulate low lift-to-drag-ratio flight experience. The technique

4. Kempel, Painter, and Thompson, "Developing and Flight Testing the HL-10," p. 29.

involved landing an F-104 "dirty," with power off and with flaps, landing gear, and speed brakes extended.[5] The pilots found it exciting to fly the F-104 this way, but they had to be careful to avoid losing control of the aircraft. The pilots' choice later on in preparing for low lift-drag flight and later for flying chase on lifting-body flights was clearly the F-104, a reliable aircraft that had the pilots' full confidence.

The F-104, as it turned out, provided excellent training experience for pilots as preparation for lifting-body flights. The aircraft's high-speed landing gear and large-speed brakes could be used to duplicate lifting-body lift-to-drag characteristics. The aspect ratio of the F-104 was only about 2.46 with a low-speed, clean configuration at a maximum lift-to-drag ratio of approximately 5.7. With the engine at idle, gear and flaps down, and modulation of speed brakes, the lift-to-drag ratio could be made to simulate each of the lifting-body configurations. In this sort of power approach at 170 knots, the lift-to-drag ratio was approximately 2.9. Thus, the lift-to-drag-ratio envelope of the F-104 essentially blanketed the lift-to-drag-ratio values of all of the lifting bodies.

Chasing lifting bodies in the F-104, however, was not totally without risk, as experienced by NASA pilot Tom McMurtry. Chasing one lifting-body flight, McMurtry inadvertently entered an uncontrolled spin. This was serious because the F-104 was not known as an aircraft that could successfully recover from a spin.

The incident occurred at 35,000 feet and 210 knots airspeed with gear down, flaps at takeoff, speed brakes out, and power at idle while McMurtry was maneuvering to join up with the lifting body. Maneuvering into position, McMurtry rolled to 45 degrees of bank and sensed the aircraft starting to slice to the right while in heavy buffet with the nose pitched up. The F-104 went into a spin. One of the other chase pilots, Gary Krier, saw what was happening and radioed McMurtry, calling for full forward stick and full forward trim. The F-104 was in a flat uncontrolled spin directly over the Edwards maintenance and modification hangar, rotating to the right at about 40 to 50 degrees per second.

The aircraft made four or five full turns before McMurtry stopped the rotation by holding full left rudder, neutral aileron, and stick and pitch trim at full nose-down. Recovery from the spin seemed very abrupt, completed at approximately 180 knots and 18,000 feet. The engine did not flame out, and the only configuration change made during the spin was the retraction of the speed brakes. McMurtry held the nose down until the F-104 reached 300 knots and then pulled out at slightly over 4G, the bottom of the pull-out occurring at 15,000 feet.

After the lifting body landed successfully, McMurtry joined the other chase aircraft in the traditional fly-by. Later, during the post-flight debriefing, discussion of

5. Gene J. Matranga and Neil A. Armstrong, ***Approach and Landing Investigation at Lift-Drag Ratios of 2 to 4 Utilizing a Straight-Wing Fighter Airplane*** (Edwards, CA: NASA TM X-31, 1959).

the lifting-body mission seemed almost trivial in comparison with McMurtry's description of his experience in the F-104.

From Analog to Digital Computer Simulation

By the mid-1960s, flight simulation had become an essential part of flight research at the Flight Research Center.[6] Even Paul Bikle, who had been somewhat skeptical of the early simulation work with the M2-F1, was beginning to recognize the importance of flight simulation in planning lifting-body flights. Over the three and half years of flight-testing the HL-10, three NASA simulation engineers—Don Bacon, Larry Caw, and Lowell Greenfield—were involved. Air Force Captains John Rampy and John Retelle were later involved with the HL-10 simulator and stability and control.

The HL-10 real-time simulator was primarily an engineering tool, not a pilot-training simulator per se. The simulator was fixed-base—that is, it had no cockpit motion. It had an instrument panel similar to that of the flight vehicle as well as a pilot's control stick and rudder pedals closely approximating those of the actual aircraft. No visual displays were available, all piloting tasks being accomplished by using the instruments. The instrument panel included indicators showing airspeed, altitude, angle of attack, normal acceleration, and control surface position. A three-axis indicator provided vehicle attitude and sideslip information.

Both engineers and pilots used the simulation extensively. Engineers used the simulator for final validation of control-system configuration. Control gearing selection was always difficult with the fixed base. The pilots wanted high sensitivity until they were airborne. Then, the simulation engineers had to decrease the gearing. Modern motion simulators of today have moving cockpits and give high fidelity to control gearing selection.

The simulation was used later to plan each research flight mission, specifying maneuvers and determining flight profiles including Mach numbers, altitudes, angles of attack, and ground track needed for mission objectives to be achieved. Emergency procedures were also practiced on the simulator, inducing various failure modes and selecting alternate landing sites. The pilots were relatively willing subjects once they knew they would be flying the actual mission, and the training paid large dividends. From this information, flight cards were assembled and distributed at crew briefings to all involved personnel, including chase and B-52 pilots, the mission controller, participating flight-research engineers, and NASA and Air Force managers. Coordination was critical to the success of each mission.

The pilots were unanimous in reporting that, once in flight, the events of the mission always seemed to progress more rapidly than they had in the simulator. As a

6. See Thompson, *At the Edge of Space*, pp. 70-71.

result, engineers and pilots experimented with speeding up the simulation's integration rates, or making the apparent time progress faster. They found that the events in actual flight seemed to occur at about the same rate as they had in the simulator once that simulation time was adjusted so that 40 simulator seconds was equal to about 60 "real" seconds. Only the final simulation planning sessions for a given flight were conducted in this way. In his book, ***At the Edge of Space***, Milt Thompson discussed how this difference between simulator seconds and seconds as perceived by pilots in actual flight was first discovered during the X-15 program, the first aircraft research program that made extensive use of simulation in flight planning and pilot training, and resolved by Jack Kolf who originated the concept of fast-time simulation, compressing simulator time to approximate time as it appeared in actual flight.[7]

The first simulation of the HL-10 was done with the Pace 231R analog computers then in use at the Flight Research Center. The real capability of the analog computer was its ability to integrate differential equations. Because the equations of motion for the lifting bodies were differential equations—as are all equations of motion for aerospace vehicles—the simulation engineers mechanized them on available analog computers. During the early to mid-1960s, digital computers were primarily used for data reduction, not for real-time simulation. Analog computers were fast, having no problems with cycle time. However, they left much to be desired when it came to mechanizing highly nonlinear functions common to aerodynamic data. Simulation engineers at the Flight Research Center could generate these nonlinear functions on analog computers—but only with great difficulty, patience, perseverance, and a lot of time.

With the aerodynamic data for the modified HL-10, the simulation engineers wanted to mechanize the highest fidelity simulation possible, so they purchased a relatively high-speed digital computer to generate the nonlinear functions. They interfaced the digital and analog computers, using the analog system for the integrations, and moved into the world of hybrid computerization. This approach proved quite successful, allowing them to make fast, efficient changes to the aerodynamic database when they were needed.

Although the program engineers were not aware of it, the simulation engineers—Don Bacon, Larry Caw, and Lowell Greenfield—decided to experiment with moving all of the mathematical computations, including the integrations, to the digital computer. Afterwards, they gave a demonstration of an all-digital, real-time computer simulation. Program engineers Bob Kempel and Wen Painter couldn't tell the difference. Neither could the pilots Bill Dana, Jerry Gentry, Pete Hoag, and John Manke.

The HL-10 program thus achieved another milestone, having successfully made the transition from simulation by analog computer to real-time simulation by digital computer. Today, analog computers have nearly gone the way of the dinosaur. At the Dryden Flight Research Center since the mid-1970s, virtually all flight simulation has been done by using high-speed digital computers.

---

7. ***Ibid.***

Brown-Bagged Panic: Crashing the Simulator

After the second flight of the HL-10 in March 1968, Jerry Gentry and John Manke alternated as pilots of the vehicle during eight more glide flights in subsonic configuration before the HL-10 was fitted with the rocket engine for supersonic flight in transonic configuration. The aerodynamics became quite different in the transonic, or "shuttlecock," configuration with the rudders moved outboard and the elevon flaps moved upward. Now that the flight envelope of the HL-10 was expanding to supersonic speeds at higher altitudes, everyone on the project was a little edgy, including the pilots.

A diligent research pilot, John Manke didn't believe in wasting time when it came to practicing on the simulator for upcoming flights. One day, to practice for his first supersonic flight with the HL-10, he showed up during the lunch hour, bringing his bagged lunch with him. No program engineers were still in the room, and Manke was left alone with the simulator once the simulation engineer left for lunch after loading a data set into the simulator. However, inadvertently, the simulation engineer had loaded the wrong data set—a demonstration set, not used for flight planning, that had directional stability set at zero.

Manke began simulated flight, unaware of the error. Achieving planned altitude for acceleration to supersonic speed, Manke pushed the nose over, toward zero angle of attack, and the vehicle became violently unstable in the lateral direction. The result? Manke "crashed" in the simulator.

To a simulation engineer, "crashing" in simulated flight may seem no big deal, for the engineer may be primarily conscious of the fact that simulated flight is not real flight, but to a pilot who uses a simulator as a pre-stage to actual flight, "crashing" in the simulator can be a major big deal. With no program engineers around at the time, Manke expressed his concerns at once to NASA management.

As a result, project engineers Bob Kempel, Berwin Kock, Gary Layton, and Wen Painter quickly found themselves in the "Bikle barrel," Bikle's wood-paneled executive office, trying to explain to Paul Bikle, Joe Weil, and several other members of the NASA management why they were trying to kill a perfectly good test pilot—a guy all the project engineers liked very much, even if he was from South Dakota.

Kempel recalls feeling a long way from the office's door as a means of escape from this very uncomfortable meeting, a formidable barrier of high-level managers standing between it and the HL-10 project engineers. Once the feeding frenzy had abated, it occurred to the project engineers that the wrong data set must have being used. They explained the problem and followed up with a demonstration in the simulation lab, showing that with the correct flight data set loaded into the simulator, no dynamic instability occurred.

From Rocket Power to Supersonic

On 23 October 1968, Jerry Gentry attempted the first lifting-body powered flight in the HL-10. Unfortunately, the rocket failed shortly after launch. Propellant was jettisoned, and an emergency landing was made successfully on Rosamond Dry Lake located about 10 miles southwest of Rogers Dry Lake within the boundary of Edwards Air Force Base. A few weeks later, on 13 November, John Manke successfully flew the HL-10 for the first time in powered flight.

Five months later, on 17 April 1969, Jerry Gentry flew the X-24A for its first flight. After the B-52 had launched Gentry in the X-24A that day, it was mated with the HL-10 and then launched John Manke in the HL-10 for that vehicle's fifteenth flight. For the first and only time in lifting-body history, two flights in two different vehicles were launched the same day from one mothership.

It's traditional, following a maiden flight, to douse the pilot. After Gentry's first flight that day in the X-24A, during the party at the Edwards Officers' Club, someone decided the swimming pool could be used for Gentry's dousing. However, no one had noticed the pool was nearly empty. Fortunately, Gentry survived his shallow immersion with only a few cracked front teeth.

A few weeks later, on a beautiful spring day in the Mojave Desert, John Manke made the world's first supersonic lifting-body flight in the HL-10 on 9 May 1969. The flight plan for the first supersonic flight of the HL-10 called for launching approximately 30 miles northeast of Edwards AFB, igniting of three rocket chambers, rotating to a 20-degree angle of attack, maintaining that angle of attack until the pitch attitude was 40 degrees, and maintaining that pitch attitude until the vehicle reached 50,000 feet. At that altitude, according to the flight plan, Manke would push over to a six-degree angle of attack and accelerate to Mach 1.08, afterwards changing angle of attack, turning off one rocket chamber, and maintaining a constant Mach number while gathering data. Landing was planned as a typical 360-degree approach with a landing on Runway 18.

Later, Manke reported that there had been no significant problems during the flight and that generally everything had gone really well. Indeed, the actual flight went almost entirely according to plan. On this historic seventeenth flight, the HL-10 actually rose to an altitude of 53,300 feet and achieved a speed of Mach 1.13, both slightly above the planning figures.

Some special engineering events preceded the first supersonic lifting-body flight. These included completely reviewing the wind-tunnel aerodynamic data and reassessing the predicted dynamic and vehicle controllability characteristics in transonic and supersonic flight regimes. Between Mach 0.9 and 1.0, the data indicated an area of low, and even slightly negative, directional stability at angles of attack of 25.5 degrees and above. Predictions and the simulator showed acceptable levels of longitudinal and lateral-directional dynamic stability at all angles of attack and Mach speeds. The engineering team also prepared a detailed technical briefing that was presented to the NASA and Air Force management teams.

The HL-10, like all lifting bodies, had very high levels of effective dihedral. This characteristic—along with positive angles of attack and acceptable levels of directional stability—ensured lateral-directional dynamic stability almost everywhere in the flight envelope. Before the flight, the HL-10 team demonstrated to project pilot John Manke that the HL-10 would exhibit this dynamic stability even if the static directional stability was zero or slightly negative, provided that the angle of attack did not approach zero.

Bob Kempel recalls that the actual flight was probably not as exciting as the events leading up to it. From what Kempel remembers of the flight, it was relatively uneventful—except for the fact of going supersonic. Nevertheless, in his book ***On the Frontier***, Richard Hallion calls this first supersonic flight "a major milestone in the entire lifting-body program," adding that "the HL-10 [later] became the fastest and highest-flying piloted lifting body ever built."[8]

Faster and Higher

About nine months after Manke's first supersonic flight, during the 34th flight of the HL-10 on 18 February 1970, Air Force pilot Major Pete Hoag bested Manke's Mach 1.13, achieving Mach 1.86. Nine days later, on the 35th flight, NASA pilot Bill Dana took the HL-10 to an altitude of 90,303 feet.

Hoag's Mach 1.86 in the HL-10 was, indeed, the fastest speed achieved in any of the lifting bodies. From B-52 launch to touchdown, the flight lasted 6.3 minutes. Except for the Mach number exceeding the preflight prediction, the flight was fairly routine.

The HL-10 had been launched about 30 miles southwest of Edwards AFB, heading 059 degrees magnetic, at 47,000 feet. According to flight plan, all four rocket chambers were ignited immediately after launch. The vehicle was rotated to a 23-degree angle of attack until a pitch attitude of 55 degrees was attained, that pitch attitude held until the vehicle reached 58,000 feet, followed by a pushover to zero G—angle of attack near zero—maintained until the fuel was exhausted. Predicted preflight Mach speed had been 1.66 at 65,000 feet. However, Hoag achieved Mach 1.86 at 67,310 feet.

The fourth NASA research pilot to fly the HL-10, Bill Dana had flown the 199th and last flight of the X-15 in late October 1968, six months later making his first HL-10 glide flight on 25 April 1969. When he took the HL-10 to 90,303 feet on 27 February 1970, Dana not only flew the HL-10 higher than it had ever been flown before, he also set the record for the highest altitude achieved by any lifting body. From B-52 launch to touchdown, the flight lasted 6.9 minutes.

---

8. Hallion, *On the Frontier*, p. 162. See immediately below in the narrative for the details Hallion is summarizing here.

Dana's flight to maximum altitude was launched under the same initial conditions as Hoag's nine days earlier, except that launch was executed 2,000 feet lower (45,000 feet) with pushover 9,000 feet higher (67,000 feet) to a seven-degree angle of attack held to Mach 1.15. Speed brakes were deployed at that altitude and speed, angle of attack then increasing to 15 degrees. According to the flight plan, maximum altitude was to have been reached at this point. What was achieved was Mach 1.314 and an altitude of 90,303 feet. The rest of the flight was fairly routine, except that touchdown was changed from Runway 18 to Runway 23 to avoid high crosswinds.

HL-10: Lift and Drag

For success, any aerospace vehicle must have adequate controllability. The modified HL-10 had very good control characteristics. Equally important to the HL-10's success in the lifting-body program was its ability to generate and control lift, plus its relatively high lift-to-drag ratio in its subsonic configuration. As measured in flight with the landing gear up, the HL-10's maximum lift-to-drag ratio was 3.6, so its best subsonic glidepath angle was approximately -16 degrees (below the horizontal reference).

The HL-10 and the M2-F2 can be compared in terms of their lift-to-drag characteristics, for although the two lifting bodies had considerably different configurations, their missions were similar. Maximum lift-to-drag ratio for the HL-10 was 14 percent higher than for the M2-F2. Although both vehicles had similar lift-curve slopes, the M2-F2 had a much lower angle of attack at a specific lift coefficient than the HL-10. Both vehicles initiated a 300-knot approach at a lift coefficient of approximately 0.15, resulting in a flight path angle of about -25 degrees for the M2-F2 and about -16 degrees for the HL-10 and a landing approach at pitch attitude of about -25 degrees (nose down) for the M2-F2 and about -8 degrees for the HL-10. (The approach flight path angle of commercial airliners in 1990, by comparison, was about -3 degrees.) Never a problem for the lifting-body pilots, the steep approaches for landing were always breath-taking to watch, especially the particularly steep descents of the M2-F2.

At about Mach 0.6, the lift-to-drag ratio of the HL-10 in transonic configuration was approximately 26 percent lower than it was in subsonic configuration. Since lowering the landing gear decreased the lift-to-drag ratio by about 25 percent, the common landing technique with the HL-10 involved flaring in the clean subsonic configuration, then lowering the landing gear in the final moments of flight.

"Dive Bomber" Landing Approaches

After its modification, the HL-10 was often rated by the pilots who flew it as the best flying lifting body in terms of turns and the "dive bomber" landing approaches typical of the lifting bodies. On a rating scale of 1 to 10, with 1 being the highest rating, the average of ratings for the HL-10 was a 2. Each pilot was asked to evaluate various piloting tasks or maneuvers during each of his flights. Following a flight, the pilot then completed a questionnaire involving numerical evaluations as well as comments. Of the 419 numerical ratings given on flights, 43 percent were 2, with 98 percent of the pilot ratings being 4 or better. The best possible rating, a 1, figured on 3 percent of the ratings, while the worst rating received, a 6, showed up in only 0.7 percent of them.

Following the modification, the HL-10 presented no serious problems in piloting. Pilots found it relatively easy to fly, the HL-10 landings being no more difficult than making a similar power-off landing approach in an F-104. To some, the steep and unpowered landing approaches seemed to be mere sport, a daring maneuver of little or no advantage. Often, until they have been apprised of the benefits, spacecraft designers and engineers have failed to appreciate the advantages of these steep, unpowered approaches. Air Force pilot Jerry Gentry, in fact, advocated this type of approach even for the F-104 in normal operations. Pilots found the high-energy, steep, unpowered approach to be safer and more accurate than the recommended low-energy approach for the F-104 because it allowed gentler, more gradual changes in altitude.

What Gentry and other pilots found to be true in the HL-10 and F-104 had been known to be true for many years in terms of accuracy in the old dive bombers, where it was generally accepted that the steeper the dive angle, the greater the accuracy. The approach task in the HL-10 involved positioning the vehicle on a flight path or dive angle to intercept a preflare aim point on the ground, similar to the targeting task of the dive bomber. The difficulty of the HL-10's task was minimized by using a relatively steep approach of -10 to -25 degrees.

There was never a problem in the HL-10 of being short on energy, because the approaches generally were begun well before the peak of the lift-to-drag curve—that is, at high speeds and relatively low angles of attack. Energy was modulated while arriving on the desired flight path by slowing, accelerating, or remaining at the same speed and using the speed brakes to make needed changes in the flight path. Speed brakes are critically important on any aircraft landing with power off, for speed brakes can be used much like a throttle to vary the parameters of the landing pattern. What is more, speed brakes add only minimal weight to the vehicle and require no fuel. The small emergency landing rockets installed on the HL-10 were used only for experimental purposes and during the first flight of the vehicle. On all later flights, the speed brakes were consistently used, instead.

Later in the lifting-body program, many spot landings were attempted in the HL-10 because it was generally believed that unpowered landings on a conventional runway would one day be a requirement, as it is currently with the Space Shuttle. On those spot landing attempts, the average miss distance was less than 250 feet. This degree of accuracy in landing is a benefit of the high-energy, steep, unpowered approach typical of the HL-10 lifting body.

Higher speed in the landing approach also provided better controllability of the vehicle. For example, a contemporary aircraft landing approach with high power and low speed was much more demanding on a pilot. During the low-speed approach to a carrier deck, the aircraft was operating past the peak of the lift-to-drag-ratio curve—that is, at a relatively high angle of attack—where the vehicle's stability, controllability, and handling qualities were degraded and where engine failure could be catastrophic.

Although the pilots thought highly of the HL-10 for its excellent control in turns and during the steep landing approaches, most of the pilots did not like the visibility they had from inside the vehicle. Even though the pilot was located far forward in the HL-10, the canopy had no conventional canopy bulge. What is more, the rails at the lowest extent of the plexiglass canopy were relatively high, providing a sideward field-of-view depression angle of approximately 16 degrees to the right and somewhat less on the left, due to the canopy defrost duct. Pilots in the HL-10 were supplied routinely with a squirt bottle of water to use in case the flow from the defrost duct wasn't enough to handle the fog of condensation obstructing their view during critical moments of flight.

The plexiglass nose window provided excellent forward vision for navigation and maneuvering for touchdown. Unfortunately, the nose window was lens-shaped and, distorting distance like the wide-angle sideview mirrors on today's cars and trucks, gave the pilots the impression that they were higher off the ground than they really were. After one of his flights in the HL-10, John Manke reported that he had touched down before he wanted to, due to the distorted view out the nose window. Some pilots on their first flights in the HL-10 waited until they were critically close to the ground before they extended the landing gear. Only the accumulation of actual flight experience in the HL-10 alleviated this problem for the pilots.

Mysterious Upsets and Turbulence Response

As one might imagine, all of the lifting bodies possessed some unique aerodynamic characteristics. One of the most unusual is what is called "dihedral effect." On conventional winged aircraft, the "dihedral" is the acute angle between the intersecting planes of the wings, usually measured from a horizontal plane. The "dihedral effect" is essentially the aerodynamic effect produced by wing dihedral that is related to the tendency of a winged aircraft to fly "wings level." It is also the effect which produces a rolling tendency proportional to the angle of sideslip (side gusts). Even though lifting bodies don't have wings, they possess very large amounts of dihedral effect, which means that a very large amount of rolling tendency is generated for small amounts of sideslip, the primary reason why lifting bodies were flown with "feet on the floor"—that is, with pilots deliberately keeping their feet off the rudder pedals. Rudder would induce sideslip, and the lifting bodies would respond primarily with rolls.

Each of the lifting bodies experienced flight through turbulence which caused pilot anxiety out of proportion to the involved "upsets," or uncommanded disturbances of unknown origin. These upsets were so different from upsets as experienced in conventional winged aircraft that the pilots frequently became disturbed when encountering any turbulence in a lifting body. Aerodynamically, the lifting bodies were significantly different from winged aircraft and one might expect them to respond quite differently to turbulence, but what we were experiencing was something entirely new and unknown.

The pilots could not agree on what particular sensations triggered their anxiety, but they said that they often felt on the verge of instability. Early in the lifting-body program, the pilots reported feeling that the vehicles were going to "uncork" on them. Once the pilots became convinced that there was no real instability and that the vehicle disturbances were caused by turbulence, they rode through the disturbances with little concern.

The gust response of an unwinged vehicle is considerably different from that of winged aircraft. In conventional aircraft, turbulence primarily affects the vertical, felt in the seat of the pants. In a lifting body, turbulence primarily affects the horizontal, producing small amounts of sideslip disturbance, resulting in a high-frequency rolling sensation. This was particularly true at lower elevations where turbulence could be most severe. Following the crash of the M2-F2 in May 1967, the pilots became even more sensitized to upsets close to the ground, the crash of the M2-F2 during landing linked to the rolling motions from such an upset that temporarily disoriented the pilot. In turbulence at low elevations, the pilots felt they might be experiencing some impending dynamic instability in the vehicle, even though the engineers assured them that they were not.

Mysterious upsets occurred at altitude as well, usually during the powered portion of a profile. The pilots found these upsets "spooky." The program engineers hypothesized that these upsets were caused by wind shears. Consequently, on one flight a movie camera was positioned on the ground directly beneath the planned ground track, since the LR-11 rocket motor always left a distinctive white trail of exhaust condensation, or contrail, in any and all atmosphere conditions. Just before launch, the upward-facing camera was turned on to record the launch, powered portion of the flight, and the pilot's radio transmissions. As the pilot flew the powered portion, he called out where the vehicle "felt squirrely" in the lateral direction. Later, playing the film showed that the vehicle had indeed encountered wind shears, as shown by the disturbed contrail, when the pilot had reported that the vehicle "felt squirrely."

Over time and with experience, the pilots came to accept that the turbulence response of the HL-10 was considerably different from that of conventional winged aircraft and that the upsets did not mean that they were on the threshold of dynamic instability. This was new territory in aerospace exploration, one in which the lifting-body pilots and engineers found themselves having to separate the real from the imagined.

Experiments with Powered Landings

After its 35th flight, when all of the major program objectives had been met, the HL-10 was reconfigured for a powered approach and landing study conducted over two flights on 11 June and 17 July 1970. For the study, the LR-11 rocket engine was removed and three small hydrogen-peroxide rockets were installed. The objective was to study shallower glide angles during final approach. Ignited during approach, the rockets reduced the angle of approach from approximately 18 to 6 degrees. The 37th and final flight of the HL-10, piloted (like the 11 June flight) by Pete Hoag, was also the last of the powered approach flights in this study.

The overall results of the study were negative, powered landings having no advantage over unpowered ones for the lifting body. Indeed, shallower powered approaches in the lifting body provided none of the benefits normally obtained in winged aircraft from powered landings. Another conclusion from the study was that the normal approach technique for any space re-entry vehicle—even if equipped with airbreathing engines with go-around capability—should be to operate the vehicle as if it were unpowered, relying on the engines only if the approach were greatly in error. This conclusion proved to be of great influence later in the design of the Space Shuttle, especially the decision not to install landing engines on the Shuttle. Yet much credit for that decision should go to Milt Thompson, especially to his perseverance in campaigning vigorously for unpowered Shuttle landings.

When we total up the flight time for the HL-10 in its 37 flights between 1966 and 1970, we come up with 3 hours, 25 minutes, and 3 seconds. Was that enough time for us to prove the value of the lifting-body concept? We think so, especially every time we watch a Space Shuttle landing.

CHAPTER 7

WINGLESS FLIGHT MATURES

Costing between two and three million dollars and involving 60 NASA employees, the rocket-powered lifting-body programs for the M2-F2, M2-F3, and HL-10 were major undertakings for the Flight Research Center. However, this effort seems small in comparison with the several hundred million dollars being invested by the United States at that time, mostly through the Air Force, in lifting re-entry technology.

In the early 1960s, the Air Force funded several studies within the aerospace industry of winged-vehicle configurations, variable-geometry slender bodies, and high-volume lifting bodies. However, having less confidence in wingless designs, the Air Force committed several hundred million dollars to winged vehicles, most of this money channeled between 1960 and 1964 into two hardware programs, the manned Boeing Dyna-Soar X-20 and the unmanned McDonnell ASSET (Aerothermodynamic/elastic Structural Systems Environmental Tests) programs.

In 1963, a major shift occurred within the Air Force regarding aerospace concepts, interest waning in the winged-vehicle concept as interest grew steadily in the concept of high-volume lifting bodies. By that time, we had had nearly a year of solid flight experience at the Flight Research Center with the M2-F1 lifting body, and I was spending most of my time developing and selling the supersonic lifting-body program to NASA management. Since November 1960, the Air Force had had the Martin Aircraft Company under contract for developing a full-scale flight-testing program of a lifting re-entry vehicle. By December 1963, Martin had selected the SV-5 configuration, following the results of wind-tunnel tests on various lifting re-entry designs.

A high-volume lifting body, the SV-5 was the brain child of Hans Multhopp, an aerodynamicist at the Martin Aircraft Company. The SV-5 quickly became the centerpiece of a new Air Force program known as START (Spacecraft Technology and Advanced Reentry Tests). Established in January 1964, START consisted of dual programs—the unpiloted PRIME (Precision Recovery Including Maneuvering Entry) and the piloted PILOT (Piloted Lowspeed Tests).

In early 1964, I visited the Martin Aircraft Company to gather information on the SV-5 and possibly gain some support from Martin and the Air Force in convincing NASA management to fund a supersonic lifting-body flight-test program. I met Hans Multhopp, introduced to me as Martin's chief scientist and the designer of the SV-5. A soft-spoken man with a heavy German accent, Multhopp seemed to be highly respected and admired by others in Martin engineering. After a conversation with him about the SV-5, I could understand why he was so highly respected, for his knowledge of aerodynamics and aircraft design was impressive.

A former aeronautical engineer, Multhopp had worked during World War II for the Focke-Wulf Flugzeugbau in Bremen, Germany, first as head of the aerodynamics department and then as chief of the advanced design bureau. One of his projects at Focke-Wulf was designing, in conjunction with Kurt Tank, the Ta-183. Information on the Ta-183 design obtained by the Russians at the end of World War II greatly influenced the design of the Russian MIG-15 jet fighter. The Pulqui-II, a derivation of the Ta-183 design flown in Argentina after World War II, had been built by former Focke-Wulf employees who had fled to Argentina.

Whisked out of Germany at the end of World War II, Multhopp went to work for the British at Farnborough. There, he designed the swept-wing British Lightning fighter, using calculation techniques he had developed. After four years, however, the British found his arrogance intolerable and he was sacked. He then became the chief scientist for the company that eventually became the giant American aviation and space contractor, Martin Marietta.

Multhopp was able to convince Martin management as well as the Air Force that the SV-5 shape was superior to NASA's M2-F3 and HL-10 shapes on the basis of six features. First, the SV-5 was a maneuverable lifting body with no essential surface components that would be destroyed on re-entry from orbit. Second, the vehicle had a hypersonic lift-to-drag ratio of 1.2 or better, permitting a lateral range of 1,000 miles. This feature would enable a recall to any preselected site at least once a day as well as emergency recall to a suitable location from every orbit.

Third, the low-speed aerodynamics of the SV-5 were suitable for making a tangential landing without resort to automatic controls. Fourth, volumetric efficiency was as high as possible, the shape giving as much volume forward as possible for center-of-gravity control. The resulting configuration gave more room up front for the pilot and equipment. The center-of-gravity could then be positioned sufficiently forward to provide adequate vehicle control without resorting to an unstable vehicle with a negative static margin. Fifth, positive camber was included in the body, allowing trimmed lift conditions at lower angles of attack as well as a high subsonic lift-to-drag ratio of about 4.0. Sixth, in regards to pilot visibility, the SV-5 cockpit canopy design was superior to that of the M2-F3 and the HL-10.

My first meeting with Hans Multhopp at Martin in early 1964 also turned out to be my last. After that visit, he seemed simply to disappear from public view. Later, when the X-24A was being flown at Edwards Air Force Base as the final stage of the PILOT portion of the SV-5 program, I was surprised to learn that my Air Force colleagues at Edwards had never even heard of Hans Multhopp. At that time, there was still considerable resentment in this country about using German engineers in American aerospace projects. Consequently, it became the usual practice to keep German engineers at low profile. However, this was not always true. A good example of an exception to this practice was Wernher von Braun, who rose to high rank in NASA in full public view and made a significant contribution to our space program.

The PRIME unpiloted SV-5 program began in November 1964. The Space Systems Division of the Air Force Systems Command gave the Martin Aircraft Company a contract to design, fabricate, and test a maneuverable re-entry vehicle in order to demonstrate whether a lifting body could, in fact, be guided from a straight course and then returned to that course. Martin had already been studying lifting re-entry vehicles for some time—the company had, after all, been in the Dyna-Soar competition—and had invested more than two million hours in lifting-body studies.

Martin Aircraft Company refined the SV-5 design into the SV-5D, an 880-pound aluminum vehicle with an ablative heat shield. The Air Force ordered four of the SV-5D aircraft, which it designated the X-23A. Between December 1966 and mid-April 1967, the Air Force launched three of these vehicles atop Atlas boosters that blasted them at 14,900 miles per hour over the Pacific Ocean Western Test Range toward Kwajalein. The three vehicles performed so well that the Air Force canceled the fourth launch to save money. The PRIME project demonstrated that a maneuvering lifting body could indeed successfully alter its flight path upon re-entry. These tests also conclusively confirmed that lifting bodies were maneuverable hypersonic re-entry configurations.

From SV-5P to X-24A

As an expansion of Martin's PRIME work, the Air Force and Martin derived PILOT—a proposed "low-speed" (Mach 2) research vehicle that the Air Force could test for its supersonic, transonic, and subsonic-to-landing behavior. Martin designated the vehicle the SV-5P.

Colonel Chuck Yeager, then commandant of the Edwards Test Pilot School, had been a fan of the lifting bodies since his flight in the M2-F1. At the time, he had told Paul Bikle that the first lifting body handled well and that he would like to have a few jet-powered versions to use for training future lifting-body pilots. After learning of what Yeager had said, Martin proposed the SV-5J, a low-speed lifting-body trainer powered by a small turbojet, for use by the Air Force's test pilot school at Edwards.

Nothing came of this proposal, although Martin built the shells for two such vehicles and even tried to entice Milt Thompson to fly the SV-5J when it was completed. NASA had no interest in the vehicle, and Thompson was committed to supporting the objectives of the NASA lifting-body program. Calculations showed that the vehicle, because of its high drag and low thrust, would not only have marginal climb performance but would actually be dangerous to fly. Nevertheless, Martin offered Milt Thompson $20,000 if, on his own time, he would simply get the vehicle airborne. Thompson offered to accept Martin's $20,000 if he could get it airborne by simply bouncing the SV-5J a few inches into the air by running it across a two-by-four on the runway. Martin didn't accept Thompson's "flight plan."

Meanwhile, the SV-5P development program was progressing smoothly. In May 1966, the Air Force gave Martin a contract for building one SV-5P. Martin began development under the direction of engineers Buz Hello and Lyman Josephs. About a year later on 11 July 1967, Martin rolled out the SV-5P at its plant in Baltimore, Maryland. The Air Force designated the vehicle the X-24A.

X-24A Crew and Wind-Tunnel Testing

Selecting staff and crew for the X-24A lifting-body project coincided with the winding down of the X-15 program. The X-24A gained experienced flight planners and flight-test engineers from both the X-15 and M2-F2 programs, including NASA's Jack Kolf and the Air Force's Johnny Armstrong, Bob Hoey, Paul Kirsten, and David Richardson. Chief NASA flight planner for the X-15, Jack Kolf became project manager of the program under the direction of John McTigue.

Norm DeMar became operations engineer for the X-24A. His crew included crew chief Jim Hankins; mechanics Chet Bergner, Mel Cox, and John "Catfish" Gordon; inspector LeRoy Barto; avionics technician Ray Kellogg; instrumentation technicians Bill Bastow and Jay Maag; and, from Martin Aircraft Company, electrical engineer Bob Moshier and hydraulics and mechanical-systems engineer Jack Riddle. Wen Painter and Sperry Rand's Ron Kotfilm worked on the vehicle's stability and augmentation system.

On 24 August 1967, the X-24A was delivered to Edwards Air Force Base. An experienced lifting-body pilot who had probed the instability boundaries of the M2-F2, Jerry Gentry was assigned as project pilot. Although Paul Bikle and Jerry Gentry were anxious to keep the X-24A on schedule, the vehicle did not fly for the better part of another two years. The vehicle was not released for program activity until 5 October 1967, when DeMar and his crew began preparing the X-24A for wind-tunnel tests at NASA Ames.

Although the X-24A left Edwards on 19 February 1968, wind-tunnel testing at NASA Ames did not begin until 27 February, the extra days at the wind tunnel used to prepare the vehicle with a removable coating to simulate the ablative roughness that would be encountered after the heat of re-entry. Roughness measurements from recovered PRIME vehicles were used in preparing this coating. Afterwards, the X-24A was wind-tunnel tested with two skin conditions, with a clean metal skin and with the rough surface stuck to the skin with an adhesive.

The rough surface seemed to cause a significant reduction in the lift-to-drag ratio for landing, a reduction that would, in turn, reduce the time available for correcting control inputs during actual landings on re-entry from space. These conclusions, along with other aspects of the wind-tunnel test data on the X-24A, were published a year later in a NASA report written by Jon S. Pyle and Lawrence C. Montoya, two engineers at the NASA Flight Research Center, entitled ***Effects of Roughness of Simulated Ablated Material on Low-Speed Performance Characteristics of a Lifting-Body Vehicle.***[1] However, flight tests at Edwards were planned for the vehicle only with a clean metal skin.

---

1. Jon S. Pyle and Lawrence C. Montoya, ***Effects of Roughness of Simulated Ablated Material on Low-Speed Performance Characteristics of a Lifting-Body Vehicle*** (Washington, DC: NASA TM S-1810, 1969).

Problems and More Problems

After the X-24A returned to Edwards on 15 March 1968, DeMar and his crew began preparing it for flight. However, problems began to appear that would slow them down. First, since the cockpit instrument panels had not been designed at Martin to be removable for check-out and maintenance, DeMar and his crew had to spend two months installing connectors on all electrical and pressure fittings in the panel. Next, when the hydraulic control system was operated, the actuators started leaking, so they had to change all of the servo valves. During hangar tests of the control system, when runaway control-surface oscillations were put in, structural feedback resulted, eventually traced to its origin through the soft actuator structural mounts. This problem occurred because Martin engineers—in their zeal to avoid having to add weight to the vehicle nose for balance—had designed the X-24A to be very light in its aft end, where the actuators were supported.

Competition had sprung up earlier between the Martin designers of the X-24A and the Northrop designers of the M2-F2, M2-F3, and HL-10. The Martin designers knew that the Northrop designers had had to add either nose ballast or redundant structure in the noses of the Northrop-built lifting bodies to maintain center-of-gravity, and they vowed that they would not do the same in their design of the X-24A lifting body. They claimed that one of the assets of the X-24A shape was that it offered more volume forward for the pilot, allowing heavy equipment to be installed in the nose. However, the Martin designers had been so frugal in weight control that the structure and actuators in the aft end of the X-24A were of minimum size and thicknesses. In fact, the aft end of the X-24A was so light that 140 pounds of ballast had to be added to it to balance the vehicle for flight.

DeMar and his crew had to beef up the structure to eliminate the control-system dynamic feedback encountered in ground tests, and this process delayed the X-24A schedule substantially. According to DeMar, he was called into Bikle's office almost weekly during this time to explain to Bikle and Gentry what was causing the latest delay. Even more delays came about as a result of the new wave of caution and conservatism that had engulfed the Flight Research Center following the crash of the M2-F2 the year before. It had always been a tradition at the Center to have a Flight Review Board made up of engineers and technicians not involved in a project to recommend when a project's aircraft was ready for flight-testing. The Board formed to examine every detail of the X-24A, however, proved to be very picky. Extra tests on systems were needed to assure the Board that the vehicle was flight-worthy, further delaying the schedule.

X-24A Glide-Flights, 1969-1970

After the X-24A finally was declared ready for flight, Jerry Gentry was set to pilot the vehicle in its first glide-flight on 17 April 1969, nearly two years after its roll-out

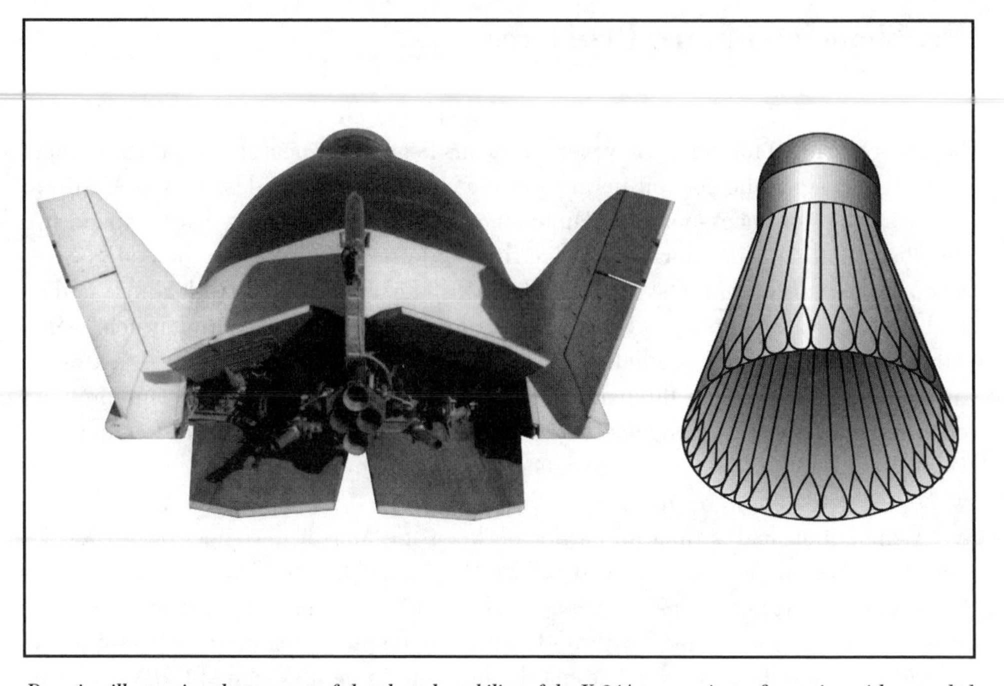

Drawing illustrating the concept of shuttlecock stability of the X-24A transonic configuration with extended control surfaces (original drawing by Dale Reed, digital version by Dryden Graphics Office).

by Martin Aircraft Company. Gentry's first glide-flight of the X-24A turned out to be almost as hair-raising and exciting as Milt Thompson's and Bruce Peterson's first glide-flights respectively of the M2-F2 and HL-10.

Milt Thompson had experienced a lateral-directional pilot-induced oscillation (PIO) in the M2-F2 at low angles of attack when he moved the manual rudder-aileron interconnect wheel the wrong way. Bruce Peterson had experienced pitch and roll oscillations in the HL-10, the result of flow separation on the outer vertical fins at high angles of attack. Because of this flow separation, the pitch and roll control in the HL-10 was ineffective. Jerry Gentry faced somewhat different problems on the first flight of the X-24A.

The X-24A was more automatic and complex than either the M2-F2 or HL-10. First of all, roll control on the X-24A could come from either the lower split flap or the upper split flap. Roll control could be shifted from the lower to the upper flap in either of two ways: by scheduling an automatic biasing (shuttlecock configuration) of the upper flap for transonic flight, or by the pilot pulling back on the stick, resulting in retraction of the lower flap and extension of the upper flap.

Eliminating the dependency on the pilot to set the rudder-aileron interconnect manually, the X-24A included an automatic system that changed the interconnect ratio as the angle of attack varied. For its first glide-flight, the X-24A was launched from the B-52 with its upper flap set at 21 degrees upward from the body's upper skin to give the lowest drag during the subsonic glide to landing. This setting meant that all roll control during the flight would come from the lower split flap.

The first flight of any air-launched lifting body is unique. With the vehicle's very rapid rate of descent, the pilot has only about two minutes to evaluate actual flight characteristics and determine that no serious deficiencies exist that could compromise a safe landing. During that same two minutes, the pilot also has to perform enough maneuvers in the aircraft to allow lift-to-drag performance and longitudinal trim to be determined, information that later will be compared with wind-tunnel predictions so that the second flight can be approached with an even higher degree of confidence.

The launch of the X-24A from the B-52 into its first glide-flight was smooth. However, one minute into the flight, the automatic interconnect system failed, causing the interconnect to stick in one position. During the landing approach at two degrees angle of attack and 300 knots, Gentry experiences an uncomfortable lateral-directional "nibbling." He said that the sensation was similar to one he had experienced in the M2-F2 with a characteristic that developed into a severe lateral-directional PIO tendency with large bank-angle excursions. At approximately 1,800 feet above ground, to stop the roll oscillation in the X-24A, Gentry increased the angle of attack to between four and five degrees, decreased airspeed to 270 knots, and used the landing rockets, a successful flare landing without rockets requiring an airspeed of 300 knots.

Just before touchdown, the lower flaps were rate-limited, the maximum surface rate from the actuators being insufficient to follow the large commands from both the roll rate-damper system and the pilot, which were in phase. During the flare, Gentry considered the longitudinal control to be good. However, due to actuator rate-limiting, the rate damper could not be fully effective during periods of surface rate-limiting. The result was that the vehicle's roll-rate excursions reached 20 degrees per second.

Something obviously needed to be changed on the X-24A. Johnny Armstrong, Bob Hoey, the NASA engineers, and the Air Force engineers Captain Charles Archie, Paul Kirsten, Major John Rampy, Captain John Retelle, and Dave Richardson analyzed the flight data and concluded that the problems with roll oscillation and elevon actuator rate-limiting were caused by the failure of the automatic interconnect system. The poor handling qualities of the X-24A during the final approach were primarily the result of the higher-than-planned rudder-to-aileron interconnect that occurred when the automatic system failed.

Once the interconnect system problem was corrected and with no other changes to the vehicle, Jerry Gentry piloted the X-24A on its second glide-flight. However, the same problem occurred, the lower flaps again becoming rate-limited on the final approach, even though the rudder-aileron system was working properly.

Before the third glide-flight of the X-24A, the program's engineers conducted a considerable investigation by simulator to define the changes needed to improve the vehicle's flying qualities on final approach. Subsequent changes made to the control system included modifying the lower-flap control horns to approximately twice the maximum surface rate, modifying the rudder-aileron interconnect schedule with angle of attack, and increasing the control-stick force gradient and stick-damping in roll. More effective rate-damping gain settings in roll and yaw were defined. Although the X-24A's response to motion in turbulence could not be duplicated adequately in the fixed-based simulator, the X-24A engineering team concluded that the effect of turbulence significantly contributed to the control problem.

Bob Hoey recalls that the most significant cause of the oscillations on the X-24A's first and second glide-flights was "an error in the prediction of the yawing-moment-due-to-aileron for the lower flap. The error was apparently caused by flow interference around the sting in the wind tunnel when the flaps were closed to nearly zero deflection. The flight data showed that the derivative was of opposite sign than predicted. Although we suspected the problem, we didn't measure this correct value until after fl[igh]t 2, when the pilot did some aileron doublets."

In retrospect, Hoey concluded, "we were lucky on fl[igh]t 1." Not only had the interconnect stuck too high at 35 percent but even more proverse was the aileron derivative. These effects were additive, Hoey said. "Later analysis showed that Gentry was well into the predicted PIO region on that approach, and his decision to slow down and use the rockets was a good one!"[2]

During Gentry's third glide-flight, he noticed considerable improvement due to the changes in the control system. However, he continued to be concerned about the vehicle's response in turbulence. Gentry did not begin to lose this concern until, during additional glide-flights, he became convinced that the motions he was sensing stemmed from "riding qualities" aggravated by turbulence rather than from any serious deficiency in handling qualities. The increased surface rates of the lower flaps, furthermore, prevented the reoccurrence of the earlier problem with rating-limiting. Nine more glides were made in the X-24A before the vehicle's first powered flight.

X-24A Powered Flights, 1970–1971

By combining much larger fuel tanks with a lighter-weight structure in the X-24A, Hans Multhopp and the other Martin designers theoretically achieved the potential for the X-24A to attain much higher speed and altitude than either the M2-F3 or the HL-10. All of the powered lifting bodies had the same type of rocket engine, the LR-11, with a maximum theoretical vacuum thrust of 8,480 pounds. In structure, the X-24A was nearly 200 pounds lighter than the HL-10 and 700 pounds lighter than the M2-F3. The X-24A also carried about 1,600 pounds more in fuel than did the HL-10 or M2-F3. Fuel-to-vehicle weight ratios for the three powered lifting bodies were 0.45 for the X-24A, 0.35 for the HL-10, and 0.33 for the M2-F3. The X-24A seemed to have the potential for breaking lifting-body speed and altitude records.

2. Typed comments of Robert G. Hoey to Dale Reed in conjunction with his technical review of the original manuscript for "Wingless Flight," Sept. 1993.

View of the X-24A showing eight retracted control surfaces on the aft end of the vehicle in its subsonic, low-drag configuration. (NASA photo E68 18769)

Bob Hoey, however, felt that the maximum speed of the X-24A would not be greater than what had been achieved already with the other powered lifting bodies. "The reference area of the X-24A was 162 sq[uare] f[ee]t," he explained, compared to 139 square feet for each of the other two powered lifting bodies, "so it was larger with more wetted area. The X-24A also required a larger wedge angle (more drag) for stability at transonic and supersonic speeds. This is a desirable feature while decelerating during an entry, but undesirable when trying to accelerate with a rocket."

Actual X-24A entry, Hoey continued, would use 50 degrees of upper flap and 10 degrees of outward flare on the rudder down to Mach 2, identical to the configuration of the PRIME vehicle that deployed a drogue chute at Mach 2. As speed decreased below Mach 2 in the X-24A, Hoey theorized, the upper flap and rudder bias would begin to program inwardly. "We used 40 degrees of upper flap and 0 rudder as our transonic/supersonic configuration," Hoey said, "a compromise in reduced shuttlecock stability in order to get lower drag and higher speed under power. Our simulation showed that we would only reach about Mach 1.7 for an optimum, full duration burn."[3]

Historic accounts including Richard P. Hallion's ***On the Frontier*** and ***The Hypersonic Revolution*** have suggested the X-24A had few, if any, negative points.

3. Hoey, comments to Reed.

However, the X-24A's high reputation rests on the fact that the vehicle was not allowed to be flown in what might have been very uncontrollable flight regimes. Hans Multhopp and his fellow designers at Martin had designed the X-24A exclusively as a re-entry vehicle. It had not been designed to perform well in other situations, including being launched from a B-52, climbing to altitude, and diving to achieve a high Mach speed during rocket burn.[4]

The X-24A had very serious angle-of-attack control limitations at transonic speeds. If the pilot increased angle of attack above about 12 degrees, he risked losing roll control due to roll-reversal boundary. If the pilot continued increasing angle of attack in the X-24A to near 20 degrees, wind-tunnel tests and simulations predicted the vehicle would depart in yaw from its intended direction due to lack of directional stability. According to these predictions, at these high angles of attack, neutral longitudinal stability also would occur. The X-24A also had a low angle-of-attack limitation, experiencing roll-reversal and pitch-instability problems at angles of attack lower than four degrees.

Nevertheless, it can be said that the X-24A had no constraints in handling or stability for an optimum, maximum speed boost profile. "Although the stability boundaries were well defined by flight test," Bob Hoey said about the X-24A, this use of flight-testing being fairly traditional by the end of the X-15 and M2 programs, "they DID NOT constrain the optimized trajectory. . . . We had adequate margins on both sides of the boundaries to safely fly an optimized trajectory."[5]

Flight research teams for the various lifting bodies always wanted their vehicle to surpass the speed and altitude records of earlier lifting bodies. The less restrictive control boundaries of the HL-10 allowed its pilots to be able to fly more optimum-powered trajectories on speed and altitude missions than were allowed for the X-24A. The X-24A team chose to see the HL-10's speed record as mainly a matter of luck, saying that on its speed mission the HL-10 climbed and accelerated at lower altitudes with a tailwind, then climbed slightly into a jet-stream headwind that increased airspeed and added about 0.2 of its speed record of Mach 1.86. Perhaps partly in jest, the HL-10 team replied that they had planned it that way and that perhaps the X-24A team ought to do the same.

Nevertheless, the wave of caution that engulfed the Flight Research Center following the M2-F2 crash affected flight-planning for several years for the lifting bodies still being flight-tested—the HL-10, the X-24A, and eventually the M3-F3. As a

---

4. Richard P. Hallion and John L. Vitelli, "The Piloted Lifting Body Demonstrators: Supersonic Predecessors to Hypersonic and Lifting Reentry," Chapter II: "The Air Force and the Lifting Body Concept," pp. 893-945, esp. p. 922 of Hallion, ed., ***The Hypersonic Revolution: Eight Case Studies in the History of Hypersonic Technolog***, 2 vols. (Wright-Patterson Air Force Base, OH: Aeronautical Systems Division, 1987), Vol. II; Hallion, ***On the Frontier***, p. 164.

5. Hoey, comments to Reed.

result, much care was taken to avoid crossing any possible out-of-control boundaries. Carefully considered restraint characterized the planning of maximum speed and altitude missions for the X-24A. Flight safety was paramount. Program objectives would be met if the rocket-powered lifting bodies, including the X-24A, could be flown at supersonic speeds near or greater than Mach 1.5 in order to test re-entry glide performance.

As objectivity prevailed, the X-24A team decided not to try to set speed and altitude records for the lifting bodies. Describing the X-24A team effort, Bob Hoey said, "We tried twice to get to the expected burnout point of 1.7 (actually 1.68 on the Flight Request). Both flights resulted in engine malfunctions. The X-24B program had already been approved, so we decided that the benefit of another tenth in Mach number was not worth the added risk to the vehicle and crew." As a result, "we stopped the X-24A program without ever flying a speed profile to burnout."[6]

Eighteen powered flights were made in the X-24A between mid-March 1970 and early June 1971. A typical X-24A powered flight lasted just under eight minutes, consisting of a two-and-a-half-minute rocket-powered flight followed by a five-minute glide to landing. The vehicle's speed envelope in Mach number was expanded in successive small steps separated at times by pauses for investigating problems affecting handling. Primary flight objectives were not met on the first five powered flights due to system failures following launch.

Flight planning and crew preparation for the powered flights took considerably more time than had been required for the glide flights. Not only was the basic flight plan more complex for powered flight, but a large number of possible deviations had to be planned and practiced in simulation. Each X-24A pilot usually spent over 20 hours in flight simulation in preparing for each flight. Furthermore, actual flight practice in the F-104 was also increased to include landing approaches to as many as five different runways. Each of the three X-24A pilots—NASA's John Manke and Air Force Majors Jerry Gentry and Cecil Powell—performed as many as 60 landing approaches during the two weeks prior to a flight.

Generally, the primary objective of each powered flight was to perform data maneuvers near the planned maximum Mach speed, and this required precise control of the profile. Consequently, data maneuvers were generally limited to the angle-of-attack range required for profile control. To prevent the possibility of large upsetting maneuvers that could compromise the profile, all data maneuvers were done with the stability augmentation system engaged. The capability for individually operating the LR-11 rocket engine's four chambers made it possible to select a reduced thrust level upon reaching the desired test conditions to provide additional data time at quasi-steady flight conditions.

On 19 March 1970, Jerry Gentry piloted the X-24A in its first powered flight, reaching well into the transonic region by achieving a speed of Mach 0.87. After we

6. Hoey, comments to Reed, underlining in the original.

analyzed the data from the first powered flight, there were two changes. First, the center-of-gravity in the X-24A was moved forward by removing 140 pounds of ballast from the tail. Second, to help reduce longitudinal control sensitivity, the upper flap was biased upward from 30 to 35 degrees above the aircraft's body surface—in effect, opening the shuttlecock.

On the flights that followed, pilots John Manke and Cecil Powell steadily expanded the X-24A's performance envelope. During these flights, to increase directional stability, the shuttlecock was increased, biasing the upper flap upward to 40 degrees above the body surface, and the rudders were moved outboard. To improve handling, we also increased yaw damper gain and the rudder-to-aileron interconnect ratio.

Exactly 23 years after Chuck Yeager's first supersonic flight, on 14 October 1970, Manke piloted the X-24A on its first flight beyond Mach 1, reaching Mach 1.19 at 67,900 feet. Less than two weeks later, Manke simulated a Space Shuttle approach and landing in the X-24A from an altitude of 71,400 feet. On 29 March 1971, Manke reached Mach 1.60 in the X-24A, its fastest research flight. However, the 28th and final research flight of the X-24A on 4 June 1971 was disappointing. Only two of the LR-11 engine's four chambers ignited, limiting the X-24A to subsonic speeds.

1971: X-24A Ready for Space

The only lifting-body configuration completely flight-tested from near-orbital speeds to subsonic landing was the PRIME (a predecessor of the X-24A, which had a slightly different configuration). The unpiloted PRIME vehicle demonstrated hypersonic maneuvering flight from Mach 24 to Mach 2.0, while the piloted X-24A demonstrated maneuvering flight from Mach 1.6 to landing. By 1971, the technology existed for initiating a rapid-turnaround, low-cost, low-risk program that could place a piloted lifting body into orbit, using a Titan II booster from the Gemini program. Had such a program come into being then, it would have resulted in the world's first lifting reentry to horizontal landing a decade before the Shuttle Orbiter.

The last flight to the moon was to occur in December 1972, leaving two complete Saturn V-Apollo systems unused. One of these rocket-and-spacecraft systems would eventually be used in a joint American/Soviet space effort, the Apollo/Soyuz orbital linkup. However, in 1971, there were still no plans for using either of the two Saturn V systems. A Northrop lifting-body engineer came up with the idea of using the vehicles for launching two lifting bodies into orbit. I thought it was a great idea. So did the NASA lifting-body project manager at the time, John McTigue.

I prepared a briefing for Wernher von Braun, then in charge of the NASA Marshall Space Flight Center in Alabama, who was visiting Paul Bikle at the Flight Research Center. The briefing was about launching two lifting-body/Saturn missions, carrying the HL-10 in the same space where the Lunar Lander had fit. The HL-10 would be modified for space flight with a heat-protective ablative coating to protect its aluminum structure. In the first mission, the HL-10 would be flown unpiloted back to earth from orbit. In the second mission, the HL-10 would be flown back by a pilot on board. Because of its maturity, the X-24A probably would have been a better choice than the HL-10.

What made the concept attractive was the proven safety of the Apollo command module that would be used by three astronauts, one of whom would be the lifting-body pilot, for the launch and climb to orbit. During the first mission, the lifting-body pilot would transfer from the Apollo capsule to the cockpit of the lifting body, conduct pre-re-entry systems checks in the lifting body, and then return to the Apollo capsule. The astronauts would then send the lifting body back to Earth unpiloted for a runway landing. Later, the astronauts would themselves return safely to Earth in the capsule via parachute.

The second mission would follow the successful completion of the first, only this time the lifting body would be flown by the astronaut/pilot back from orbit for a runway landing. If the in-orbit cockpit checks of the lifting body proved to be unacceptable, the astronaut pilot could then simply return to earth with the other two astronauts in the Apollo capsule, as done in the first mission.

In my presentation to von Braun, I used a large Saturn V-Apollo model that I had built from a commercially available plastic kit. I had substituted a model of the HL-10 for the Lunar Lander module and had even devised a model of an extraction arm for placing the lifting body in free orbit. There was enough room in the model for either the M2-F3 or the X-24A, had we chosen one of those vehicles for the mission. However, at the time, I had decided to use a scale model of the HL-10 to show the compatibility of the Saturn V-Apollo with existing lifting bodies.

Wernher von Braun thought it was a fantastic idea. He told Bikle he would prepare the rockets at NASA Marshall if Bikle would prepare a lifting body at the Flight Research Center by adding an ablative heat shield to protect the vehicle's aluminum structure from the heat of re-entry. Imagine how I felt at that moment, if you will. I was sitting in a room with two of my heroes, making plans for the first piloted lifting re-entry from space—many years before the Shuttle.

Of course, I was disappointed when Paul Bikle said "no" to the project, even though I could respect why he had made that decision. He felt my idea was good, but he also believed it was a project beyond his experience and interest. Space was beyond his realm, and he was interested only in aircraft. Paul Bikle and Wernher von Braun had each demonstrated the ability to work outside the bureaucratic process. Together, I had little doubt, they would have made the proposed project a success. And if they had, we might have been able to keep the momentum going in the lifting-body program—all the way to space.

Although we still had another five years of flight evaluation to come on the M2-F3 and X-24B lifting bodies, putting a piloted lifting body into orbit would have been a fitting conclusion to our first seven years of lifting-body flight research. However, voices of support for the Space Shuttle concept were already being heard, voices that all too soon became loud enough to drown out our vocalized advocacy for the lifting-body approach in space applications. Nonetheless, our efforts in the lifting-body program had two very significant influences on the immediate future in terms of spacecraft. First, we established the concept of horizontal landing as feasible for spacecraft recovery. Second, we established the fact that landing unpowered spacecraft with gliding lift-to-drag ratios as low as 3.0 could be conducted safely and routinely.

1969: Shuttle Concept Emerges

It wasn't until 1969—after six years of lifting-body flight at the Flight Research Center—that NASA's top-rank decision-makers and planners decided to switch from parachute recoveries of piloted spacecraft to horizontal landings. Chief of engineering at NASA Johnson Space Center, Max Faget was one of the leading figures who, at the time, was still hanging on to the parachute concept in spacecraft recovery. In fact, it was in 1969, while he was promoting the "Big G" concept for building a big Gemini capsule that could carry 12 astronauts, that he became convinced that the concept of horizontal landing was good and immediately switched sides. Studies began at NASA Johnson Space Center on lifting bodies, delta-wing configurations, and a straight-wing vehicle with a conventional horizontal and vertical tail designed by Max Faget himself. Studies led by Gene Love at NASA Langley evaluated candidates for the Shuttle configuration.

Lifting bodies remained major contenders for the Shuttle configuration until two significant events took place in 1969. The first was the invention of the lightweight ceramic tile. The second was the mandate by Congress that the Shuttle design satisfy Air Force as well as NASA requirements, including the Air Force's requirements for hypersonic lift-to-drag ratio and a full-access payload compartment about the size of a railroad boxcar.

The early ablator heat shields, developed for spacecraft such as the Apollo capsule, could be applied directly to lifting bodies with much less weight penalty than when applied to winged vehicles. However, with the invention of the lightweight ceramic tile by Lockheed Space Systems (later improved by Howard Goldstein and his team at NASA Ames), winged vehicles constructed of such low-cost materials as aluminum could compete with the lifting bodies as candidates for space. Thin surfaces, such as those found on wings and tails, could be covered with the tiles, adding only minimum weight. Minimum use of the heavier newly-developed carbon-carbon tiles could also protect leading-edge high-heat areas of winged vehicles.

Even though NASA had been granted the responsibility for developing the Shuttle, Congress dictated to NASA that the Shuttle design also had to satisfy requirements of the Air Force, which called for a payload size and cross-range requirements roughly twice those of NASA. The typical hypersonic lift-to-drag ratios of the high-volume lifting bodies that we were flight-testing were between 1.2 and 1.5, which would have served any of the projected NASA missions for hauling people and cargo to and from orbit. However, the Air Force projected greater cross-range capability requiring hypersonic lift-to-drag ratios as high as 2.0, a requirement that made winged vehicles more attractive as Shuttle candidates.

The payload requirement of the Air Force was about 50,000 pounds to low orbit, to be contained in a compartment roughly 15 by 60 feet, or about the size of a railroad boxcar. Easy access to this compartment also required the use of full-size doors that could be opened in space. This requirement narrowed down the potential spacecraft shape to what basically resembled a rectangular box with lifting surfaces (wings and tails) attached to it, plus a rounded nose on the front and rocket motors on the back. Two basic shapes evolved for final consideration: Max Faget's configuration with unswept wing and tail surfaces, and a delta-wing design with a vertical tail attached. Studies continued through 1972, when NASA selected the delta-wing shape for the Shuttle.

Phoenix Rising: From M2-F2 to M2-F3

In the winter of 1970, two powered lifting bodies were in the air over Edwards Air Force Base and a third would enter flight testing by early June. Very popular with the pilots after its modification, the HL-10 was flown more times than any other of the rocket-powered lifting bodies, its final flight occurring 17 July 1970. Since the spring of 1968, it had been flown 36 times by four pilots—10 times by John Manke, nine times each by Bill Dana and Jerry Gentry, and eight times by Pete Hoag. The X-24A was about halfway through its two-year flight-test program by the spring of 1970, ultimately being flown 28 times—13 times by Jerry Gentry, 12 by John Manke, and 3 by Cecil Powell.

The M2-F1 next to Shuttle prototype, Enterprise, showing the comparative sizes of the two vehicles. The Space Shuttle with its delta wings was selected over a lifting-body shape for the first reusable launch vehicle, but later the X-33 employed a lifting-body configuration. (NASA photo EC81 16288)

Like the mythic phoenix rising from its own ashes, the M2-F3 emerged from the wrecked M2-F2 after a nearly three-year "inspection process" of the M2-F2 that had crashed on 10 May 1967. Working closely with the Northrop lifting-body crew in Los Angeles, John McTigue parlayed resources from the shops at the Flight Research Center and Northrop along with about $700,000 from NASA Headquarters for this "inspection process," and by 2 June 1970, an essentially new rocket-powered lifting body—the M2-F3—was ready for its first flight.

McTigue had kept costs down for the M2-F3 by means of several methods. For instance, he appropriated idle X-15 crews during the winter months when the X-15s could not fly because the normally dry lakebeds used for landing experimental aircraft were wet. He also had sheet-metal and machined structural parts made in NASA's shops to Northrop's drawings and specifications and then sent to Northrop's Hawthorne facility for assembly, a uniquely cooperative venture between a government agency and a contractor that involved a most cost-effective use of labor and facilities, keeping expenses to an absolute minimum. McTigue also had the full support of Paul Bikle, a man with a reputation for supporting thrifty approaches in flight research.

Working under the direction of McTigue, Meryl DeGeer had kept the original M2-F2 team intact and involved in the building of the M2-F3. The original M2-F2 crew chief, Bill LePage, and mechanics Jay King and Bill Szuwalski continued on with the M2-F3. Although the M2-F3 resembled the M2-F2 externally, several systems had been modified, relocated, or added. The four-chamber LR-11 rocket engine, for example, was turned on its side so the lower flap could be retracted without having to build a bulge into the shape of the M2-F3's lower flap. Furthermore, heavier items were moved forward and lighter items were moved aft to help eliminate nose ballast used in controlling center-of-gravity.

Some people consider the M2-F3 the "purest" lifting-body configuration, for it had no horizontal projections or tail surfaces that could be considered small wings of some sort. The other lifting bodies had canted fins projecting into horizontal and vertical planes. By 1970, we became convinced that any engineering information that we could produce from M2-F3 flight tests would be very valuable to those designing future spacecraft. Consequently, we decided to use the M2-F3 for conducting control-system research.

The first lifting body, the lightweight M2-F1, had used a very basic mechanical control system of pushrods and cables moved solely by the pilot's muscles. There were no power systems such as hydraulics or electric actuators because the pilot didn't need them to fly the M2-F1. Only when the heavyweight lifting bodies came along—starting with the M2-F2 which, fully fueled, weighed nearly ten times as much as the M2-F1—were hydraulic controls necessary to help the pilot move the control systems against the high pressures caused by high-speed flight.

The high speeds of the heavyweight lifting bodies introduced another control problem common to all the heavyweights, the tendency for overshoot or oscillation when the pilot made a control input. Although this problem tended to manifest itself in varying degrees, depending on the configuration, it arose from the high inertia (mass distribution) and low aerodynamic damping of these vehicles. To solve this problem, we added a rate-damping stability augmentation system (SAS) to all of the heavyweight lifting bodies.

Figuratively speaking, the SAS worked like a very fast secondary electronic or robot pilot that shared control with the slower human pilot. Either "pilot" could move the control surface independently. However, since many of our early stability augmentation systems were single-string or nonredundant, we never allowed the "robot" to have more than 50 percent of the authority on the control system, not trusting it to have more control than the human pilot. We also limited the control authority mechanically to guarantee that the pilot would always have 50 percent or more control in case of electronic failure in the SAS.

The primary task of the SAS was to respond to rate gyros by telling the control surfaces to oppose angular rate movements. We called this process "rate damping" because it slowed or resisted motions of the lifting body. Often, the SAS would oppose the pilot's control inputs, telling the control surfaces to move in the opposite direction to slow down the vehicle motions commanded by the pilot. To keep the pilot and the SAS from this kind of conflict, we designed a special washout circuit for the SAS, allowing the pilot to make normal—but not high-rate—turns.

In transforming the M2-F2 into the M2-F3, we used the basic mechanical portion of the M2-F2's control system. However, we increased authority in the speed brake, modifying the rudders to allow 25 degrees of outboard deflection, and increased aileron deflection from 10 to 20 degrees. Yaw was controlled through the rudder pedals that deflected either of the two rudder surfaces on the outboard side of the two outer vertical fins.

The primary manual control system in the M2-F3 was an irreversible dual hydraulic system. Pitch was controlled by moving the center stick longitudinally, positioning the lower flap. Roll was controlled by moving the same stick laterally, differentially positioning the upper flaps.

By adding a center fin to the M2-F3, we gained true roll control with differential body flaps, no longer having the vehicle's nose moving in the opposite direction from adverse aileron yaw, as had happened with the M2-F2. In essence, we eliminated roll reversal. Even though simulation with the new wind-tunnel data told us that the rudder-aileron interconnect was not needed once the center fin was added to the M2-F3, we kept the manual interconnect control wheel in the cockpit in case we wanted to use it during the flight-test program.

Two vehicle configurations—subsonic and transonic—were used to provide adequate stability at transonic speeds and low drag (that is, an increased lift-to-drag ratio) for approach and landing. For shuttlecock stability at speeds higher than Mach 0.65, the upper body flaps were moved from the average position of 11.8 degrees to 20 degrees upward from the body surface. Outboard biasing of both rudders was used solely as a speed brake—not for transonic shuttlecock stability, as was the case for the HL-10 and X-24A.

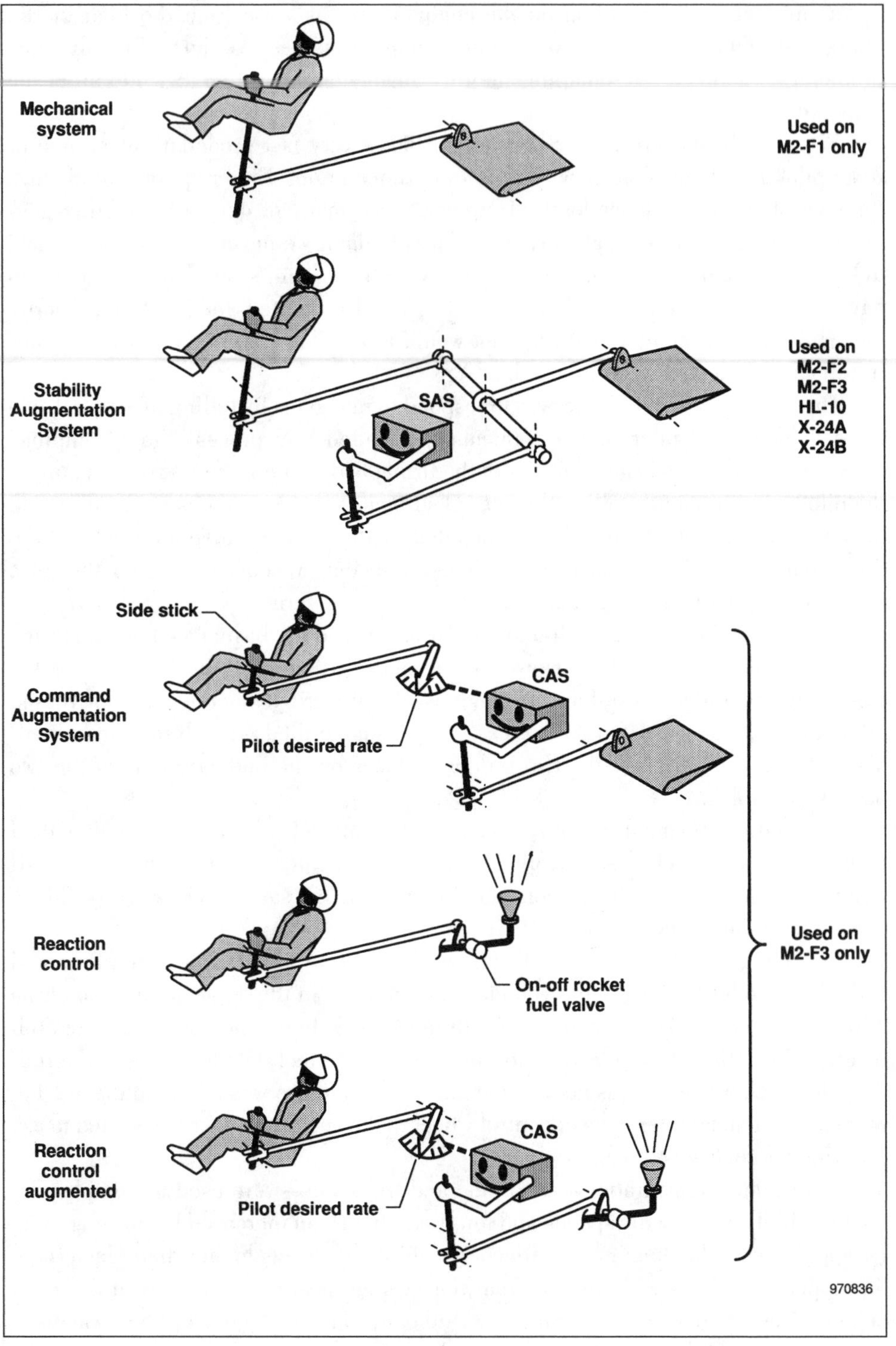

Schematic showing how the lifting-body control systems differed. The M2-F3 was a test-bed for four different control systems including reaction controls (original drawing by Dale Reed, digital version by Dryden Graphics Office).

Three Experimental Control Systems Added

To add reliability and flexibility to the electronic portion of the control system in the M2-F3, we eventually replaced the original single-string SAS of the M2-F2 with a triply-redundant Sperry electronic control system similar to the one used in the X-24A. We also added three new control systems to the M2-F3, supplemental to the basic system, using simple rate-damping controlled by the pilot's center stick. We used a second "sidearm" control stick in the cockpit for flying the M2-F3 with the three different experimental control systems. The pilot could always revert back to the basic center-stick control system by throwing a switch on the center stick or on the front panel. We planned to install these systems after the speed and altitude envelope for the M2-F3 had been expanded while using the basic center stick and SAS.

The first experimental control system for the M2-F3 was a rate command augmentation system (CAS) in the roll and pitch axis, which we hoped would improve pilot control by smoothing out the configuration's nonlinear longitudinal aerodynamic characteristics. Conceptually, the CAS differed from the SAS normally used in the lifting bodies. Instead of sharing control of the control surface with the pilot and being capable of independent operation, as was the case with the SAS, the CAS received instructions from the pilot's control stick and then used gyro and other information to actually fly the vehicle for the pilot. For instance, if the pilot wanted a certain pitch or roll rate, he would move the side-stick accordingly. After receiving the electrical signals from the pilot's side-stick, the CAS would fly the M2-F3, avoiding overshoots and oscillations and steadying the vehicle at the angular rate the pilot had indicated by stick position.

With lead and lag electronic networks, the CAS actually could do a better job than the pilot in flying a dynamically unstable aircraft. In 1970, however, we didn't trust electronics the way we do today. We gave the CAS only 50-percent authority, the pilot retaining 50-percent mechanical authority so the pilot could control the M2-F3 with the center stick if a failure occurred in the CAS. Today, high-speed aircraft routinely use command augmentation systems with 100-percent authority that are based on redundant digital computers.

We added another feature to the CAS for maintaining a pilot-indicated angle of attack. The CAS would maintain constant angle of attack when the cockpit hold switch was engaged if the pilot's side-stick was in centered position. The side-stick had a detent switch so that when it was out of center, angle-of-attack hold was disengaged and a rate dependent on stick position was commanded until a new angle of attack was reached. Centering the stick would engage angle-of-attack hold. When angle-of-attack hold wasn't desired, the pilot could turn it off with the cockpit switch and then only rate command was operative. Another switch on the side-stick provided a vernier so that angle of attack could be changed without taking the stick out of detent. The pilot could regain center-stick control with the SAS at any time by disengaging the CAS switch on the instrument panel or on the center stick.

The second experimental system installed in the M2-F3, a reaction-control rocket system, offered potential weight reduction and simplified design for potential future lifting-body re-entry vehicles. Such a vehicle could be greatly simplified if the same rockets used for maneuvering in space could be used for control during landing. Four 90-pound-thrust hydrogen-peroxide rocket motors installed on the base of the M2-F3 were designed to be operated in pairs, providing either rolling or pitching moments for roll and pitch control.

The rockets could be operated only in two states—basic ON-OFF (or "bang-bang")—with no capability for variable thrust. Effective rolling or pitching moments could be achieved only by pulsing the rockets' burn times to produce the desired impulse for changing the vehicle's motion. At first, a pilot operated the rocket system by using a spring-loaded toggle switch on the right console. Later, we replaced the toggle switch with a side-arm controller—obtained surplus from an old World War II formation stick—that enabled the pilot to use his right wrist rather than his fingers to operate the system to produce the necessary "beep-beep" or "bang-bang" motion.

A third experimental system installed in the M2-F3 was a CAS to control the reaction-control rocket system. The ON-OFF scheme of controlling the rockets seemed crude and marginal, so the CAS was modified to control the rockets rather than the aerodynamic control surfaces.

June 1970: Bomb on the Ramp!

The flight-test program for the M2-F3 benefited from the experience gained in the M2-F2, HL-10, and X-24A flight-test programs. Meryl DeGeer served as operations engineer only through the first two glide flights of the M2-F3 and then was reassigned as operations engineer on the newly established YF-12A flight program at the Flight Research Center that involved three of the Lockheed Blackbirds, similar to the SR-71A reconnaissance aircraft. Herb Anderson, who had been operations engineer on the HL-10 through its last flight, took over as M2-F3 operations engineer.

Most of the time, preparations for the M2-F3 flight tests progressed smoothly, methodically, and safely. However, an extremely dangerous incident occurred in June 1970 as the M2-F3 was being prepared for a powered flight following four unpowered glides. While hanging under the B-52's wing, the M2-F3 was being fueled on the ramp. During the fueling operation, crew member Danny Garrabrant noticed liquid was spilling out of the liquid oxygen vent onto the ramp.

Normally during fueling, the liquid-oxygen tank and the water-alcohol fuel tanks in the M2-F3 and other lifting bodies were protected by a "quad valve," a dual-redundant check valve that keeps the fuel from flowing into the liquid-oxygen tank. However, both sides of the valve failed on this occasion, allowing the fuel and liquid oxygen to mix, something that had never happened with any of the other lifting bodies. The mixture in the tanks immediately froze due to the temperature of the liquid oxygen (-270 to -290 degrees Fahrenheit), creating a bomb. The slightest jar could set off a gigantic explosion on the ramp under the fully fueled B-52.

M2-F3 launched from B-52. (NASA photo EC71 2774)

At once, Garrabrant sounded the alarm to his crew chief, Bill LePage. Herb Anderson and LePage immediately alerted the Air Force. The area was evacuated. All flights at Edwards Air Force Base were canceled, including all supersonic over-flights, for the jar from a sonic boom could trigger the explosion.

Anderson and LePage then set out to defuse their bomb. Using padded tools and being extremely careful not to drop anything on the M2-F3's tanks, they eventually succeeded, but only after several very long hours of extreme danger to themselves and the aircraft.

Flight-Testing the M2-F3

Project pilot Bill Dana flew the M2-F3 on 19 of its 27 flight missions, including the first three of four glide flights for determining how its characteristics compared with those of its predecessor, the M2-F2. Even though he had not flown the M2-F2 since 1967, Jerry Gentry piloted the M2-F3 on his final lifting body flight in February 1971. Two other pilots made the other 7 of the M2-F3's powered flights—four by John Manke and three by Cecil Powell.

After the end of the vehicle's flight-test program in late 1972, Bill Dana helped write a pilots' report on the flight characteristics of the M2-F3 that included not only his own observations but also those of John Manke, Cecil Powell, and Jerry Gentry. Published in 1975, this final NASA report on the vehicle's handling qualities entitled

"Flight Evaluation of the M2-F3 Lifting Body Handling Qualities at Mach Numbers from 0.30 to 1.61" was written by Bob Kempel and Alex Sim as well as Bill Dana.[7] This report was based on the pilot ratings for all flights and is the main source for the comments, details, and summarized results that follow.

Beginning 2 June 1970 and ending 16 December 1971, the first 13 of the 27 flight tests were made using only the vehicle's center-stick system, with in-flight maneuvers to evaluate control characteristics with the SAS on and off. Maximum Mach speed for these flights was 1.27. After the thirteenth flight, the M2-F3 was grounded for six months—until July 1972—while the experimental control systems and side-stick were installed for evaluation during the final 14 flights. The last flight occurred on 20 December 1972, the M2-F3 during the course of its flight-test program achieving a maximum speed of Mach 1.61 and altitude of 71,500 feet.

Glide Flights and Landings

During the first half of its flights, in glide and at subsonic speeds, the M2-F3 flew very well with the SAS on. Adding the center fin had made a dramatic change in the configuration, transforming the "angry machine" of the original M2-F2 into the very controllable and pleasant-to-fly M2-F3. The pilots reported that control in both longitudinal and lateral-directional axes was excellent with the rate-damping system (SAS) on. While the M2-F3 proved it could also be flown during glides with the SAS turned off in all axes, vehicle response was very sensitive and the pilots had to exercise great care to keep from over-controlling in both longitudinal and roll axes. According to the pilots, without the SAS, the M2-F3's nose would "hunt" up and down and roll maneuvers were "jerky."

During landings from the glide flights, the M2-F3 demonstrated characteristics that distinguished it from the other lifting bodies. Of the three lifting-body shapes tested, the M2 possessed the lowest subsonic lift-to-drag ratio. This fact did not create traffic-pattern difficulties due to the careful planning that went into each flight to provide sufficient altitude for comfortable landing under both normal and emergency conditions.

The low lift-to-drag ratio, however, did require more of the pilot's attention on final approach and flare than had been needed with the HL-10. Flare speed varied from 260 to 320 knots, but 260 knots proved insufficient to hold the aircraft off the ground while "feeling for the runway." About 290 knots of preflare airspeed gave a reasonable float time. However, the faster the final approach, the more comfortable it was for the pilot. Flare altitude also had to be carefully monitored for the vehicle to come level just above the ground, varying between 600 feet for final approach at 260 knots to 100 feet for 320-knot approaches.

---

7. Robert W. Kempel, William H. Dana, and Alex G. Sim, "Flight Evaluation of the M2-F3 Lifting Body Handling Qualities at Mach Numbers from 0.30 to 1.61" (Washington, DC: NASA Technical Note D-8027, 1975).

Turbulence response in the M2-F3 resembled that of the HL-10 and X-24A. A side gust would cause a high-frequency roll oscillation that would damp out without pilot input, the type of response caused by the vehicle's excessively high amount of effective dihedral. At first, low-level turbulence would make the M2-F3 pilots apprehensive due to the unusual nature of the vehicle's response. As with the HL-10 and X-24A, however, their apprehension decreased as additional experience showed that the unusual response did not mean the vehicle was on the threshold of divergent lateral oscillation. Nevertheless, we chose not to fly the M2-F3 on days when we expected high turbulence in the atmosphere over Edwards.

Having made sixteen of the X-15 flights, including its last flight, Bill Dana tended to be disappointed with the M2-F3's speed brakes. Spoiled by the X-15's powerful speed brakes, he wasn't impressed with the lesser effectiveness of those on the M2-F3. Dana also did not like the vehicle's large nose-down pitching moment when the speed brakes were applied by outboard biasing of both rudders.

The flat upper deck of the M2-F3 challenged the pilots' visibility, requiring them to switch back and forth quickly between looking over the top side and looking down through the nose window at their feet. The biggest problem with visibility in the M2-F3 was visually judging altitude just before touchdown when the nose was at high angle. Historic accounts claim that fighter pilots during World War II adapted well when they had little or no forward visibility due to the long noses on that era's aircraft, compensating by using their peripheral vision. Using the nose window, especially during approaches to touchdown, the pilots of the M2-F3 adjusted just as successfully to limited forward visibility.

Rocket-Powered Flight

During the vehicle's first rocket-powered flight in late 1970, Bill Dana achieved the transonic speed of Mach 0.81. However, indications appeared shortly after launch that the M2-F3 had longitudinal problems transonically. Angle of attack drifted nearly uncontrollably due to a decrease in pitch stability and changes in trim as the Mach number increased.

As speeds were gradually increased on each additional rocket-powered flight, the pilots discovered that the most longitudinal instability occurred near Mach 0.85, when they had difficulty controlling angle of attack. The center-of-gravity was moved forward with ballast added to the nose. Increasing the pitch damper gain, or sensitivity, to its maximum value also helped the pilot steady the vehicle. However, even with these changes, longitudinal stability (pitch control) was only marginally acceptable in the transonic speed range. Consequently, the longitudinal rate-damping system (SAS) was never turned off in this speed range.

In contrast, the pilots rated the roll control of the vehicle at transonic speeds as very good. Just as at subsonic speeds, the M2-F3 could be flown with the roll and yaw damping system (SAS) turned off. However, as it had been in glide flights, the vehicle was very sensitive to roll control, and the pilots had to exercise great care to avoid over-controlling the M2-F3 at transonic speeds with the roll and yaw damping system turned off.

At speeds between Mach 1.0 and 1.6, longitudinal control with the rate-damping system turned on was considerably better than it had been in transonic flight. However, the longitudinal control still wasn't as good as it was at subsonic speeds. We decided, consequently, that the longitudinal rate-damping system (SAS) should not be turned off at supersonic speeds. On the other hand, at supersonic speeds, the pilots felt comfortable about turning off the lateral-directional rate-damping system (SAS), for roll control was sensitive with this system operating and pilots had to be very cautious to avoid over-controlling in roll.

After the side-stick and experimental control systems were installed in 1972, the final 14 flights of the M2-F3 evaluated them. Generally, the pilots were disappointed in the Command Augmentation System (CAS). Bill Dana had hoped the CAS would improve the vehicle's handling characteristics at transonic speed during the rocket-burning phase. While the CAS did improve the longitudinal control in rate-command mode slightly, it was far from satisfactory. The pilots preferred not to use the angle-of-attack-hold mode, for it did not work well. Furthermore, the CAS did nothing to improve lateral control, already good using only the basic SAS. It seems we had cut costs too much in developing the CAS and had failed to optimize its potential.

The sidearm controller selected for use with the CAS proved to be too rudimentary. One spring in the side-stick provided both force gradient and breakout force. Adjusting one required great care to prevent varying the other. Changing either parameter required disassembling the stick, threatening the integrity of the assembly. We should have located or developed an electric sidearm controller with external and independently adjustable force gradients and breakouts.

The potential for improvement in the CAS was never fully achieved due to the poor physical characteristics of the side-stick plus the system's requirement that the pilot wear a pressure suit, which not only limited mobility but also aggravated the negative effects of the side-stick. Nevertheless, the potential for the CAS was recognized. In spite of its drawbacks, the system was a welcomed addition to the M2-F3.

The ON-OFF, or "bang-bang," rocket reaction-control system was first tried in roll with poor results. Manual control of the rockets was too responsive, resulting in jerky flying. Longitudinal control was not even tried for fear of losing control of the M2-F3.

In the reaction-control system with CAS, the pilot's side-stick was a proportional control with the stick's position commanding an angular roll rate. Tested in flight, the CAS responded to pilots' input command, firing the control rockets with pulses timed to give the desired results in changing or holding the vehicle's angular rate. The system worked beautifully without moving the aerodynamic control surfaces. Bill Dana rated the system as excellent. The system's quality reflected the level of achievement possible from applying experience with previous systems first developed at the Flight Research Center back in the days before the NACA became NASA, experience that was then applied on the rocket-boosted F-104 zoom aircraft and even later on the X-15 and Lunar Landing Research Vehicle.

A refinement on this rocket-control system eliminated unwanted yaw moments when applying roll control. The system worked almost perfectly in this mode when rockets were needed only to change roll rates. In the longitudinal mode, however, excessive use of the rockets was needed when the vehicle got out of trim by adjusting the longitudinal aerodynamic flap. A further refinement of the system, had we had the time and money to implement it, would have been to combine the longitudinal reaction-control rockets with the body's longitudinal flap in an automatic control system. The M2-F3 flight-test program was almost over, and we were nearly out of money. So we took what we had learned from the M2-F3, wrote our technical reports, and left the potential for application of what we had learned in the hands of the designers of future spacecraft.[8]

8. Kempel, Dana, and Sim, "Flight Evaluation of the M2-F3;" and Alex G. Sim, "Flight-Determined Stability and Control Characteristics of the M2-F3 Lifting Body Vehicle" (Washington, DC: NASA Technical Note D-7511, 1973).

CHAPTER 8

LIFTING-BODY RACEHORSES

By 1969–1970, the lifting-body program had become a major activity at the NASA Flight Research Center, Ames, and Langley. The Air Force Flight Test Center was vigorously supporting the flight-test part of the program for the M2-F3, HL-10, and X-24A. However, I was becoming concerned that a disproportionate amount of our effort was going into supporting only one type of lifting body.

The M2-F3, HL-10, and X-24A were configurations with high volumetric efficiencies, best suited for shuttle-type missions in deploying satellites and in carrying cargo and people to and from earth orbit. All three had hypersonic lift-to-drag ratios between 1.0 and 1.4, permitting a potential cross-range capability of 700 to 1,000 miles—that is, they could range from 350 to 500 miles to either side of the orbital path during re-entry. They also had adequate lift-to-drag ratios for landing.

To me, the M2-F3, HL-10, and X-24A were the lifting-body "plow-horses," and I was becoming interested in a different kind of lifting body, a class of vehicles I considered the "racehorses." The shapes of these lifting bodies had high fineness ratios with long pointed noses and flat bottoms. The more efficient of these shapes had hypersonic lift-to-drag ratios as high as 3.0, allowing a total cross range of 3,000—the ability to range 1,500 miles to either side of the orbital path. A hypersonic vehicle with a lift-to-drag ratio greater than 3.0, of course, would be considered at the top of its class in performance.

The "racehorse" class of lifting bodies could be used for special missions where flexibility was required, being able to land anywhere on earth on short notice. However, the slender shapes would not lend themselves to serving as efficient cargo containers. While these vehicles would have high aerodynamic efficiencies at hypersonic speeds, they wouldn't perform well at landing speeds and likely would need some sort of deployable wings to land.

Two of these "racehorse" shapes were the Hyper III developed by NASA Langley and the FDL-7 developed by the Air Force Wright Flight Dynamics Laboratory in Dayton, Ohio. There is some question about whether the Hyper III and the FDL-7 were true lifting-body configurations since they both had small deployable wings used for landing. Both can be called special forms of the lifting body, however, since the small wings would be stowed during most of the projected re-entry flight before landing. Another of the lifting-body features that each possessed was that, even with the wings deployed, the body still dominated the aerodynamics of the total configuration.

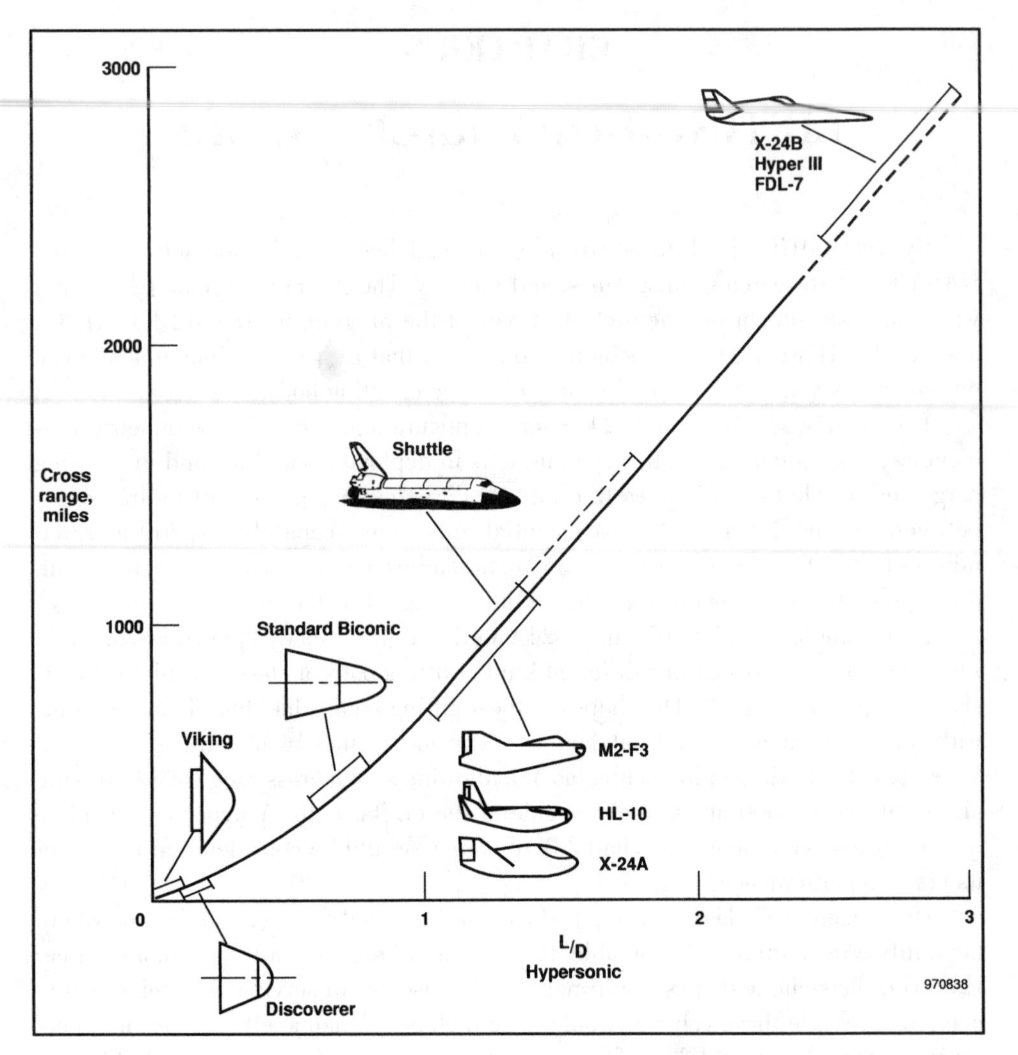

Graph showing cross range distances in miles plotted against hypersonic lift over drag for several vehicles returning from orbit. Notice that the "race-horse" vehicles such as the X-24B and Hyper III have the greatest cross-range capability—around 2,500 miles.

Model-Testing of Lifting-Body Spacecraft

By 1969, I was outside the mainstream of the on-going lifting-body program at the Flight Research Center, busy looking at new concepts and projects further into the future. Using the excellent radio-control equipment then becoming available to model-airplane hobbyists, I teamed up with Dick Eldredge to conduct several experiments in flying models of experimental spacecraft. We worked with what was called the "de-coupled mode" in which the basic re-entry vehicle is flown down to a certain point and then converted to a landing configuration by deploying either a gliding parachute or wings of some sort.

Eldredge had been the first research engineer to join my M2-F1 lifting-body team seven years earlier, and I still thought of him as my "little buddy." Although we had remained in contact with each other throughout the lifting-body buildup program, since the early days of the M2-F1 we had not had the time to brainstorm together about new ideas. This situation began to change after I got out of management with the lifting-body program in 1965 and, by 1969, I was free to think about new ideas again.

Over the years I have often compared the relationship between me and Dick Eldredge with that of the Wright Brothers. I thought of Eldredge as being a sort of brother off whom I could bounce ideas and from whom I could get constructive feedback, much as the Wright Brothers did between themselves during the first part of their career. Even the progressive changes in our careers bore some resemblance to those experienced by Orville and Wilbur Wright. At first, the Wright Brothers treated aeronautics as a hobby and had fun. All innovation begun early in their career stopped, however, once they became businessmen and project managers. They had no time for experimentation or research once they entered competition with Glen Curtiss and others and became involved with legally protecting their wing-warping and other patents. By that time, aviation was no longer fun for the Wright Brothers. It had become serious business.

I have noticed that the same changes often occur within the careers of many innovative individuals who are motivated by fun as well as the satisfaction they receive from creating something that has never existed before. When these people enter the business world, they often become unhappy, their productivity diminishing. I believe I made the right decision when I took Paul Bikle's advice in 1965 and got out of management with the lifting-body program. When I returned to engineering, I essentially returned to the realm of innovation.

As I learned from my own experience over the years, NASA Headquarters operates in such a way that priority and attention tend more easily to be given to large and costly projects. Experiments or projects by two people or a small group generally do not fit into the scheme of things at NASA Headquarters. In fact, until a project has been supervised by NASA Headquarters, pondered for some time there, and then officially blessed, it usually is not considered important by headquarters people.

Nevertheless, the small projects that result from brainstorming at the NASA centers are often exciting for those who originate them and literally love the work they do. I don't think, on the other hand, that most managers at NASA Headquarters trust those who have too much fun while working. In fact, these managers coined the term "hobby-shop projects" for referring disparagingly to projects originating outside of the mainstream and control of the master plan.

Dick Eldredge and I, however, were intrigued with the idea of doing the first flight-testing of a sleek "racehorse" configuration with a pointed nose, a design we believed would give superior performance at hypersonic speeds. As we continued our radio-controlled model flying of lifting-body spacecraft, we tested models of both the "racehorse" Hyper III and the "plow-horse" M2-F2, using a Rogallo Limp Wing gliding parachute for recovery.

We also designed and built a special twin-engined, 14-foot model mothership for carrying the lifting-body models to altitude and launching them, much as was being done with the B-52 for the full-scale lifting bodies. We envisioned future space missions where there might be a need to use the vehicle's hypersonic lateral cross-range capability to reach a meadow in Alaska, for example, and land the vehicle there softly and slowly by means of a gliding parachute for some covert military mission. Our imaginations also came up with a mission that used the hypersonic lateral range of the vehicle to take an injured astronaut back to earth, landing in a field near the hospital best able to provide the care needed.

One of us would fly the mothership by radio-control with the lifting-body model attached to its belly with a hook activated by remote control. The other would take charge of the experimental lifting body, flying it after air-launch on its own aerodynamics, then controlling it through steering control lines to a landing on its gliding parachute.

We found that the Hyper III's extremely low lift-to-drag ratio of 2.5 made it impractical to land without either a gliding parachute or deployable wings. We experimented with three types of deployable wings for the Hyper III. The first was a pair of switchblade wings that pivoted out of slots in the lower part of the body. The second was a one-piece wing that pivoted in the center and was stowed in the upper portion of the body, the right half of the wing exiting from a forward slot on one side and the left half exiting from a rearward slot on the other side. With this second type of wing, after rotating 90 degrees, the final configuration for landing was a straight wing mounted high on the body.

The third type of wing we tried was the Princeton Sailwing that had been tested in the NASA Langley full-scale wind-tunnel on a conventional glider fuselage. The Sailwing involved two D-shaped spars stowed in two slots in the body and deployed like a switchblade wing, with trailing edge cables pulling taut from a tip rib and stretching upper and lower fabric membranes from the spar to the cable. The fabric surfaces would then curve upward, like sails on a boat, forming a cambered airfoil and producing positive-lift airflow over the wing.

Hyper III and Parawing

Our second type of deployable wing—the one-piece pivoted design—proved to be the best of the three for actual flight. NASA Langley conducted wind-tunnel tests on the Hyper III without the wing up to Mach 4.6, followed by tests with the wing deployed at subsonic speeds. I put together a plan for building a full-scale vehicle at low wing-loading similar to the M2-F1. However, I proposed to fly it without a pilot onboard. The idea of flying unpiloted vehicles at the Flight Research Center was unpopular, especially with the pilots. Paul Bikle would approve the plan only if I would build the vehicle so that a cockpit could be installed for a pilot to fly it after the initial tests were completed. Later, an X-15 type of canopy would be added slightly forward of the wing to balance the piloted version.

In spite of the success of the on-going rocket-powered lifting-body program, NASA Headquarters still was not tolerant of programs as small as that of the original lifting body, the M2-F1. For this reason, I was very interested in developing a flight-test approach with the pilot doing the early hazardous flight tests in a simulator-type cockpit on the ground. This approach would put us in a better position later for getting approval for the more expensive piloted flight tests.

I managed to convince Paul Bikle that this approach had merit and we ought to give it a try. However, the idea went over like a lead balloon with the pilots. In the end, I had to turn once again to Milt Thompson for help. Even though Thompson had retired from flying, he was intrigued with the idea and offered to fly the Hyper III from a ground-based cockpit.

By this stage in 1969, I had two projects developing at the same time. The gliding parachute tests that Dick Eldredge and I had been doing with spacecraft models had attracted the interest of the NASA Johnson Space Center. I discussed our use of the limp Rogallo parachute in recovering spacecraft models with Max Faget, Johnson's director of engineering who had played a major role in designing crewed spacecraft starting with Project Mercury.[1] Not yet accepting horizontal landing as appropriate for the next space mission, Faget at the time was still backing gliding parachute concepts such as the "Big G," a twelve-astronaut version of the Gemini space capsule with one astronaut steering the capsule to flare and landing at a ground site.

While talking with Faget, I offered to develop a one-pilot test vehicle that could be launched from a helicopter and used to test a pilot's ability to fly the vehicle while looking through the viewing ports typical of spacecraft. I suggested we fly the vehicle at first by radio-control with just a dummy onboard until it was determined to be safe to fly. Faget just happened to have a borrowed Navy SH-3 helicopter that was being used to practice fishing Apollo astronauts out of the water. Enthusiastic about my idea, Faget offered to let us have the helicopter for a month, plus enough money to buy large-sized Rogallo Parawings for the project.

Paul Bikle approved our Parawing Project, as it was called, and assigned NASA pilot Hugh Jackson to it. Although he was the new kid on the block among the other NASA pilots, Jackson was considered the resident expert in parachuting, having parachuted four or five times. At best, Jackson was lukewarm about participating in the Parawing Project. He likely accepted the assignment because he wasn't yet allowed to fly the NASA research aircraft.

Dick Eldredge designed the vehicle for the Parawing Project. It was built in the shops at the NASA Flight Research Center. Since we were experienced scroungers and recyclers by this time, we used surplus energy struts from the Apollo couches in the vehicle to soften the load on the pilot in hard landings. The M2-F2 launch adapter not being used with the B-52, we used its pneumatic hook-release system to launch

1. See Henry C. Dethloff, *Suddenly Tomorrow Came . . . A History of the Johnson Space Center* (Washington, DC: NASA SP-4307, 1993), esp. pp. 62–65.

the vehicle from the side of the SH-3 helicopter. For the test configuration, we used a generic lifting-body ogive shape with Gemini viewing ports. We attached landing skids with energy straps to an internal aluminum structure containing the pilot's Apollo couch. A general-aviation auto-pilot servo was used to pull down on the parachute control lines. The pilot used a small electric side-stick to control the servo.

The plan was that before putting a pilot onboard, we would launch the lifting-body with the dummy in the pilot's seat off the side of the helicopter, deploy the parachute, then steer the vehicle to the ground by radio-control, using model-airplane servos to move the pilot's control stick. We even tied the dummy's hands in its lap so it would not interfere with the control stick. Measured accelerations in the dummy and on the airframe were transmitted to the ground to record shock loads as the parachute opened

Hyper III with single-piece, pivot wing installed. Flexible Princeton sailwing is on the ground to be installed for future tests (never performed), and one of the fabricators of the Hyper III, Daniel C. Garrabrant, is standing next to it. (NASA photo E69 20464)

and the vehicle made ground contact. By moving the pilot's stick directly with the radio-controlled servos, we qualified the entire control system downstream of the pilot's control stick.

Dick Eldredge stayed with the Parawing Project until the system had been qualified for piloted flight following 30 successful radio-controlled flights. Hugh Jackson was getting ready to make his first flight in the vehicle when the NASA Johnson Space Center decided that the next piloted space program would not make use of a gliding parachute system but would use a horizontal-landing spacecraft instead. I think Jackson was relieved when he heard this news that made his flight unnecessary. A few months later, he left the pilots office at the NASA Flight Research Center.

Hyper III Team

Meanwhile, Dick Fischer had himself assigned as the operations engineer on the Hyper III. Fischer had other aircraft obligations, but his bosses agreed to the assignment after I had accepted the decision of management at the Flight Research Center that the Hyper III program would be conducted on a low-priority basis. A long-time friend of mine and a model-airplane flying buddy, Dick Fischer was also an excellent designer of home-built aircraft who restored antique aircraft in his spare time.

Together, Fischer and I recruited Bill "Pete" Peterson, a control-system engineer on the X-15 program, to help design the control system for the Hyper III. Peterson had worked earlier for Honeywell in Minneapolis, designing the adaptive control system for the X-15. As a Honeywell employee, he had come initially to the Flight Research Center during the X-15 flight tests to help NASA with operating the X-15's control system. He was then hired by the Flight Research Center to continue working with the control system on the X-15 and other aircraft. Peterson managed to find time to help us with the Hyper III, despite the fact that he was involved with four other aircraft at the Flight Research Center at the time.

On the Hyper III, I managed to use volunteers in the same way I had originally with the M2-F1, thanks to the influence of Paul Bikle. As in the days of the M2-F1, we found that NASA supervisors were tolerant when engineers such as Pete Peterson wanted to work on volunteer projects like the M2-F1 or Hyper III and could do so while still meeting their obligations on assigned projects.

Dick Fischer designed the structure of the Hyper III, and the vehicle was built in the NASA shops. When finished, it was 35 feet long and 20 feet wide at the tail surfaces. The fuselage was basically a Dacron-covered steel-tube frame, the nose was made out of molded fiberglass, and the four tail surfaces were constructed of aluminum sheet-metal. The aluminum wing was built from the wing kit for an HP-11 sailplane.

Frank McDonald cut and fitted the steel-tube body, and Howard Curtis did the welding. NASA aircraft craftsman Daniel "Danny" Garrabrant—a highly skilled builder of model aircraft and of home-built wooden and aluminum full-size sailplanes—assembled the wing for the Hyper III. LaVern Kelly assembled the vehicle's sheet-metal tail surfaces.

Many of the people who worked on the M2-F1 worked as well on the Hyper III, including aircraft inspector Ed Browne and painter Billy Shuler. We worked closely with the NASA fabrication shops to get the Hyper III structure completed without interfering with the shops' work on other, prioritized projects. With the X-15 program winding down, I managed to recruit even more talented volunteers to work part-time on the Hyper III, including crew chief Herman Dorr and mechanics Willard Dives, Bill Mersereau, and Herb Scott.

Our skills in scrounging and recycling came in handy in building the control system for the Hyper III, which was composed of an uplink from a Kraft model-airplane radio-control system. The control surface on each of the two elevons was driven by a surplus miniature hydraulic system from the Air Force's PRIME lifting-body program. The hydraulic system was a battery-driven pump that had run two actuators for the elevons on the PRIME vehicle.

Peterson cleverly designed the system to operate from either of two Kraft receivers, depending on the strength of the radio signal at the top or bottom of the Hyper III, one receiver mounted on the top and the other mounted on the bottom of the vehicle. If either receiver malfunctioned or picked up a bad signal, an electronic circuit switched to the other receiver. Signals from the operating receiver controlled the two elevon surfaces driven by hydraulic actuators. A talented hydraulics engineer, Keith Anderson modified the PRIME hydraulic actuator system for the Hyper III.

In case we lost control during the flight tests, we mounted an emergency parachute-recovery system in the base of the vehicle. It consisted of a drogue chute that fired aft, extracting a cluster of three paratrooper-type chutes that would lower the vehicle onto its landing gear. The Northrop support contract still in effect, I managed to get the help of Northrop's Dave Gold for a few weeks. A top parachute designer, Gold had done most of the detailed design of the parachute system used on the Apollo spacecraft. Gold and John Rifenberry from the NASA pilots' life-support shop worked steadily for two weeks at the sewing machines in Rifenberry's shop while completing the vehicle's parachute-recovery system. The Flight Research Center's expert on pyrotechnics, Chester Bergner assumed the responsibility for the drogue firing system.

We tested the emergency parachute-recovery system by putting the Hyper III on a flatbed truck and firing the drogue extraction system while we were racing across the dry lakebed, but a weak link kept the three main parachutes from jerking the Hyper III off the truck. We then tested the clustered main chute by attaching it to a weight that equaled that of the Hyper III and dropping it from a helicopter. We were very fortunate that the emergency parachute-recovery system never needed to be used.

With the help of Don Yount as instrumentation engineer and Chuck Bailey and Jim Duffield as instrumentation technicians, a 12-channel FM/FM down-link telemetering system recorded data and drove instruments in the ground cockpit. Assembled by Tom McAlister, the ground cockpit was made out of plywood and looked somewhat like a Roman chariot when it was hauled out to the landing site on a two-wheeled trailer. The instruments in the ground cockpit were identical to those in our fixed-base simulator. In the center of the display, an artificial-horizon ball indicated roll, pitch, heading, and sideslip. Other instruments in the ground cockpit showed air speed, altitude, angle of attack, and control-surface positions.

First Flight of the Hyper III

Bruce Peterson piloted the borrowed Navy SH-3 helicopter that towed the Hyper III aloft for its first flight on 12 December 1969. A Marine Corps pilot before joining NASA, Peterson continued to fly helicopters and jet fighters in the Marine reserves on a restricted basis following the M2-F2 crash that cost him his vision in one eye. After the crash, Peterson also flew support aircraft during various NASA flight-research missions, although he was not allowed to fly the actual research aircraft. The first flight of the Hyper III was the last lifting-body mission in which Bruce Peterson and Milt Thompson would directly participate.

After liftoff, with the Hyper III attached to the helicopter at the end of a 400-foot cable, Peterson had a difficult time getting the Hyper III to track straight on the end of the tow-line. Afterwards, we realized that we should have installed a small drag chute on the Hyper III that could have been jettisoned after launch. As Peterson struggled to get the vehicle to track straight, Milt Thompson sat in the ground cockpit located beside the planned landing site on the lakebed, relaxed and smoking a cigarette.

After starting and stopping forward flight several times during the climb, Peterson eventually got the Hyper III to stabilize in a forward climb. When Peterson radioed that he was ready to launch, Thompson flipped his cigarette onto the lakebed and hunched over the controls, intently ready to fly the Hyper III. Peterson towed the Hyper III to 10,000 feet above the dry lakebed, where the Hyper III was released from the tow-line by an electric cargo hook. For this first flight, the Hyper III was flown with the wing fixed in deployed position, the configuration that would be flown in a final low-speed approach and landing after re-entry from space.

Peterson dropped the Hyper III in forward flight on a downwind path with a northerly heading, Thompson controlling the Hyper III from the ground cockpit. Thompson flew the vehicle in a glide three miles north, guided it into a 180-degree turn to the left, and then began steering it the three miles to the planned landing site. During the straight portions of the flight, Thompson had performed research doublet and oscillation maneuvers so we could extract aerodynamic data following the flight.

Since Thompson was flying strictly by instrument flight rules in the ground cockpit with his head down, I asked Gary Layton in the control room at the Flight Research Center to watch a radar plot board and guide Thompson by radio to landing position. Layton had often helped lifting-body pilots in this way in the past as they steered to landing sites on the lakebed runways. Since we had no experience yet in landing unpiloted vehicles at the Flight Research Center with the use of onboard video, we had not installed a forward-looking video camera in the Hyper III. Dick Fischer was standing beside Thompson in the ground cockpit to take control of the Hyper III just before the landing flare, using the model-airplane radio-control system's box during the landing-flare maneuver to touchdown.

Although the sky over Edwards Air Force Base is often clear, on this particular day in December, the sky was hazy with moisture. While the Hyper III could be seen from the ground cockpit when it was overhead, it could not be seen through the haze when it was slanted at an angle three miles away. Without visual contact with the Hyper III, Fischer had to rely on Thompson's comments to know how the vehicle was flying, Thompson steadily watching the gauges in the cockpit.

On the final approach to landing, with Thompson calling out altitudes, Fischer strained to see the Hyper III through the haze. As the Hyper III broke through the haze at about 1,000 feet, Dick said, "I see it!" Thompson replied, "You've got it!" and switched control to Fischer's model-airplane control box.

Noticing no response from the vehicle as control was transferred, Fischer deliberately input a roll to verify that he indeed had control before he executed the landing flare. Still monitoring the gauges, Thompson told Fischer that the vehicle was rolling left, and Fischer replied that, yes, he had commanded it to roll. Now certain that the vehicle was responding to his control, Fischer used the pencil-sized control sticks on his box to bring the Hyper III level and complete the flare to a soft landing. The Hyper III slid safely to a stop on its three skids, landing on the lakebed in front of Fischer and Thompson in the ground cockpit.

We were gratified by the successful first flight of the Hyper III, having gotten the flight scheduled at our last possible opportunity for using the SH-3 helicopter before it was returned the next day to the Navy. Later, as quoted in a paper that I presented at an AIAA conference, Thompson described his experience flying the Hyper III from the ground cockpit.

"During my first attempts to change the vehicle's heading," Thompson said, "the vehicle appeared to be marginally stable or even unstable in roll. Vehicle motions in response to roll-control input seemed to be erratic and much too rapid when compared to the simulation. When faced with a situation of this type in a flight or in a simulator, I have always found the best procedure is to let go of the control stick momentarily to determine whether the vehicle is inherently stable. The Hyper III motions damped immediately after the stick was released, indicating adequate levels of stability and damping. I had simply been over-controlling and exciting a pilot-induced oscillation. The over-controlling resulted from much higher roll-control effectiveness than had been predicted."[2]

The lift-to-drag ratio of the Hyper III turned out to be lower than expected. Rather than 5.0 maximum, it proved to be 4. Thompson had had to stretch the glide as much as possible to bring the Hyper III close enough for Dick Fischer to be able to see it and land it. Twice, as Thompson pointed out, the flight had shown that a research pilot

2. R. Dale Reed, "RPRVs: The First and Future Flights," *Astronautics & Aeronautics* 12 (April 1974): 31–32.

could use actual flight experience to compensate for significant deficiencies in or departures from predicted aerodynamic characteristics.

Before the flight, Thompson had worried that the lack of motion cues, particularly during short-period motions of the vehicle, might hurt his performance in piloting the Hyper III from the ground cockpit. "This apprehension was quickly dispelled once the vehicle was launched," Thompson said. "It seemed very natural to fly the gauges, just as in the simulator, and respond to what I saw rather than what I felt."

What Thompson found surprising were his reactions during the flight. "I was really stimulated emotionally and physically, just as in actual first flights," he said. After noting that he had made the first flights in such "strange vehicles" as the Paresev and the M2-F1, he said, "Flying the Hyper III from a ground cockpit was just as dramatic."

In explaining how the experience differed from flight simulators, Thompson said, "I have flown many different simulators with and without motion and visual cues, including centrifuge and airborne simulators. Although some provided a lot of realism, none stimulated me emotionally. I always knew I could hit the reset button, or in the airborne simulators, turn the vehicle back to the conventional testbed aircraft characteristics. There was no question with the Hyper III. I, and only I, had to fly it down to the landing location."

According to Milt Thompson, his experience in flying the Hyper III "tends to confirm the theory that responsibility rather than fear for personal safety is the real driver of physiological response."[3]

NASA Headquarters Says "No" to Hyper III Piloted Flights

Our single flight of the Hyper III produced good aerodynamic data and demonstrated that the vehicle was safe to fly. By early 1970, I had located in Arizona the ideal aircraft for launching the Hyper III in a piloted flight program, an Air Force Albatross SA-116B seaplane with low flight time that had never been in the water, had no corrosion, and was in excellent condition. The aircraft was available to NASA as Air Force surplus. The Albatross had sufficient structure, control authority, and performance capability for carrying the Hyper III aloft under its wing at the 2,000-pound drop-tank location for air launch at 15,000 to 20,000 feet.

Paul Bikle asked NASA Headquarters to substitute the Albatross for the C-47 currently in use at the Flight Research Center as a utility aircraft. Trading the C-47 for the Albatross on a one-to-one basis would involve no additional cost to aircraft operation at the Flight Research Center. We could also make better use of the Albatross than the C-47, for only the Albatross could serve a dual purpose, being used as a utility aircraft when it wasn't being used in air launches.

3. Quotations from the preceding four paragraphs all in Reed, "RPRVs," p. 32.

Retired M2-F1, Hyper III, and remote control models on display. (NASA photo EC70 2450)

Bikle's request was turned down. By 1970, NASA Headquarters was caught up in the throes of internal politics, flexing its muscles as it cut authority within the various NASA centers for planning their own research. Without a launch vehicle such as the Albatross, the Hyper III would never achieve piloted flight. In this way, the Hyper III fell victim to political currents within NASA Headquarters.

The Hyper III program had three strikes against it. First, it was too low-cost to get the attention and support of NASA Headquarters. Second, it had been flight-tested as an unpiloted vehicle first, taking away some of the luster it would otherwise have had if first flown as a piloted vehicle. Third, it was a variable-geometry configuration, making it less competitive in weight and complexity than the simpler lifting-body configurations.

Paul Bikle was very upset when NASA Headquarters rejected his request for the Albatross. He saw the Albatross as a tool for the Flight Research Center and, as the director of the Center, he felt he should be able to select his own tools, especially when a tool was not going to cost NASA extra money. At the time, I think he was also seeing the writing on the wall, sensing that he no longer fit in the more bureaucratic NASA of the 1970s. It was only about a year later—on 31 May 1971—that Paul Bikle retired from NASA.

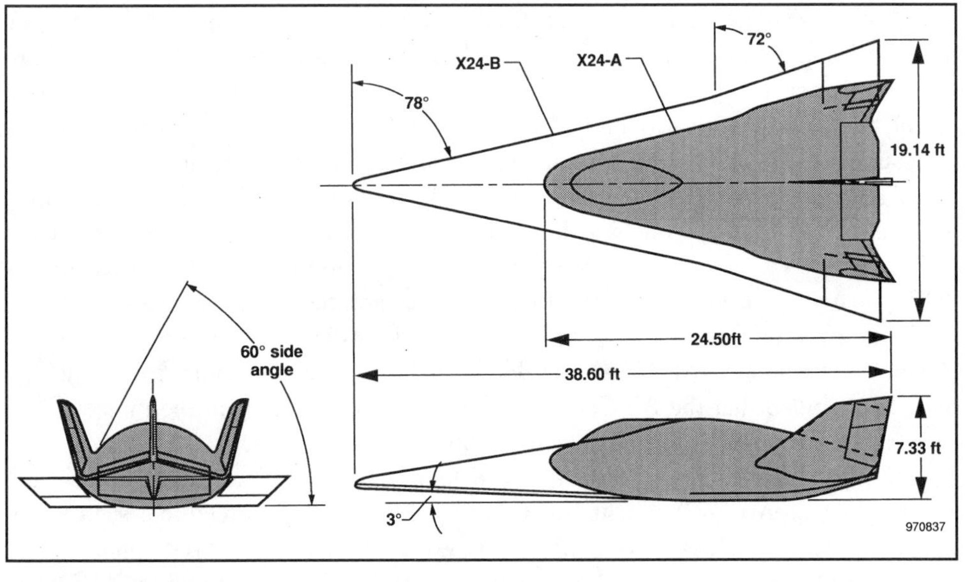

Schematic showing the X-24A conversion to the X-24B. This was a cost-saving approach to use the same systems for both configurations.

A Racehorse of Another Color: the X-24B

While we were still involved with the Hyper III, Alfred Draper and others at the Air Force Flight Dynamics Laboratory in Ohio had come up with an idea for recycling the X-24A by wrapping a new shape around it. They found that the new configuration could achieve hypersonic lift-to-drag ratios near 2.5, putting it into the same "race-horse" category of lifting body as the Hyper III which, before its wing was deployed for landing, had a hypersonic lift-to-drag ratio near 3.0. The other lifting bodies—the M-2, HL-10, and X-24A—had hypersonic lift-to-drag ratios between 1.2 and 1.4.

A distinct advantage over the Hyper III was that the new X-24A wrap-around-shape designated the FDL-8 could achieve a landing lift-to-drag ratio of at least 4.0 without variable geometry. The more slender shape of the Hyper III gave it the higher hypersonic lift-to-drag ratio of the two lifting-body shapes. However, the Hyper III had a landing lift-to-drag ratio near 2.0, making it necessary to use a deployable wing to bring the vehicle's subsonic lift-to-drag ratio up to near 4.0 for landing.

Al Draper and his colleagues at the Air Force Flight Dynamics Laboratory believed that flat-bottomed pointed shapes like the FDL-8 would prove to be useful not only for sustained hypersonic-cruise aircraft using air-breathing propulsion but also for unpowered boost-glide orbital re-entry vehicles capable of landing at virtually any convenient airfield. Furthermore, the long flat under-surface of the FDL-8 would make an ideal compression ramp for the inlet of a future supersonic combustion ramjet engine operating at speeds up to Mach 8.

At Edwards, NASA director Paul Bikle and Bob Hoey, manager of the Air Force's lifting-body program, endorsed the idea. Always attuned to thrift, Bikle was in favor of ideas that saved government money by getting the most research out of each dollar spent, the same reason why he had readily endorsed my ideas for saving money by recycling rocket engines and sharing launch aircraft with other programs.

At this point, a critical stumbling block appeared. Major General Paul T. Cooper, chief of research and technology development for the Air Force, rejected the idea of using the X-24A as a basis for the test shape that would later be designated the X-24B. Clearly opposed to the entire flight-test concept, he asked that the proposal be reviewed by a joint panel of the Air Force Scientific Advisory Board and the National Academy of Sciences. Al Draper and Bob Hoey briefed the panel on the concept. The panel concluded that the Air Force could not afford to do without the project. Thus securely endorsed, the plan advanced rapidly.

By the end of August 1970, the directors of both the NASA Flight Research Center and the Air Force Flight Test Center at Edwards had agreed that such a program was worthwhile. However, Air Force Systems Command delayed approving the program until suitable arrangements had been made for joint funding by NASA and the Air Force. Paul Bikle asked John McTigue to work with Fred DeMerritte at NASA Headquarters to come up with the money needed to get the program started. Thanks to the teamwork of McTigue and DeMerritte, NASA transferred $550,000 on 11 March 1971 to the Air Force to initiate acquisition of the aircraft. The Air Force pledged a similar amount. Finally, on 21 April 1971, the director of laboratories for Air Force Systems Command approved the program. On 4 June 1971, the X-24A completed its last flight.

On 1 January 1972, the Air Force awarded the modification contract to the Martin Marietta Corporation. A month later, on 4 February, Grant L. Hansen, the Air Force's assistant secretary of research and development, and John S. Forster, Jr., the director of defense research and engineering, signed a memorandum of understanding between the Air Force and NASA on conducting the X-24B program as a joint Air Force/NASA lifting-body venture. The memorandum was also signed by George M. Carr, NASA's deputy administrator, and Roy P. Jackson, NASA's associate administrator for advanced research and technology. The memorandum marks the official beginning of the X-24B program. Modifying the X-24A into the X-24B meant that the new research aircraft would cost only $1.1 million. The same vehicle, built from scratch, might have cost $5 million.

At the Air Force's Arnold Engineering Development Center, hypersonic wind-tunnel tests on a model of the X-24B indicated that the proposed shape performed well at those speeds. As usual, the big question was what would happen to performance when the vehicle decelerated to much lower velocities. Many, including Fred DeMerritte, expected surprises as the vehicle passed through the transonic range.

X-24B as delivered to Edwards. Notice that the original X-24A is completely disguised inside of the X-24B shape. (NASA photo E73 25283)

X-24B Shell Arrives at Edwards

On 24 October 1972, the X-24B shell built around the structure of the X-24A arrived at Edwards Air Force Base, delivered by Martin Marietta's Denver plant. Systems for the X-24B were delivered separately. The structure had grown an additional 10 feet in span and 14.5 feet in length. It weighed 13,800 pounds at launch, the X-24A having weighed approximately 12,000 pounds. The X-24B had a 78-degree double-delta planform for good center-of-gravity control, a boat-tail for favorable subsonic lift-to-drag ratio, a flat bottom, and a sloping three-degree nose ramp for hypersonic trim. The sides of the forebody aft of the canopy were sloped 60 degrees relative to the Y-plane (lateral, or left-to-right, axis).

The aerodynamic design features of the X-24B were quite distinct from those of the X-24A. Like the earlier lifting bodies, however, the X-24B also used several off-the-shelf components. Portions of its landing gear, control system, and ejection system came from the Northrop T-38, the Lockheed F-104, the Martin B-57, the Grumman F11F, the Convair F-106, and the North American X-15. It had an LR-11 rocket engine and Bell Aerosystem landing rockets.

Although the basic systems in the X-24B were the same as those in the X-24A, several upgrades and additions were made in the propulsion system, control system, and nose landing gear. The LR-11 rocket engine was modified, the vacuum thrust increased from 8,480 to 9,800 pounds by increasing chamber pressure and adding nozzle extensions. The engine started at a lower thrust level with thrust increased after the engine was stabilized.

Two outboard ailerons were added to the eight control surfaces that had been on the X-24A. The HL-10 also had ten control surfaces (the subject of the standing joke that HL-10 stood for "Hinge Line Ten"). The two new control surfaces on the X-24B were used only for roll control with a plus or minus five-degree pitch bias feature. Unlike the X-24A, the X-24B's split upper and lower body flaps were not used for roll control. The pitch control and shuttlecock biasing of the X-24B, however, were the same as on the X-24A. The triply-redundant rate-damping system used in the X-24A was retained in all three axes on the X-24B with variable gain control by the pilot. Most of the other control system features of the X-24B, including the hydraulic power supply and rudder biasing linked to the body flap biasing for transonic stability, were the same as on the X-24A. The biasing on the X-24B could also be used by the pilot for speed brake control.

Basically the same automatic aileron-rudder interconnect system was used in both the X-24B and the X-24A, although the system in the X-24B had more flexibility in operation. The amount of interconnect was automatically programmed as a function of angle of attack. As in the X-24A, the pilot could select two interconnect angle-of-attack schedules, a high-gain one for transonic-supersonic conditions and a lower-gain one for control at subsonic speeds. The pilot could also use a manual interconnect mode as backup to the automatic scheduling or for special test maneuvers.

The X-24B retained the T-38 main landing gear that had been used in the X-24A. However, unlike the X-24A, the X-24B used a modified Grumman F-11F-1F nose gear. The combination resulted in an unusual arrangement of landing gear, similar to but not as extreme as that on the X-15. The main gear on the X-24B was significantly aft of the landing center of gravity, and the three-point attitude was nose low. The landing gear was a quick-acting (approximately 1.5 seconds) pneumatic system. The main gear deployed forward, the nose gear aft, minimizing not only the movement in the center of gravity but also the change in longitudinal trim. From the cockpit, the pilot could actuate the landing gear only to the down position.

While the cockpit controls and instruments were basically the same in both the X-24A and X-24B, the X-24B alone had an F-104 stick-shaker. The shaker actuated at 16 degrees angle of attack to warn the pilot that the vehicle was approaching an area of reduced pitch stability. Later in the X-24B flight-test program, to provide additional sideslip monitoring for the pilot, an audio sideslip warning system was added.

X-24B Team: Preparing for Flight Tests

Following the end of the X-24A flight-test program on 4 June 1971, the X-24A crew, led by operations engineer Norm DeMar, was disbanded for 16 months while Martin Marietta was contracted and the X-24A was transformed into the X-24B. During this time between the disbanding of the X-24A crew and the formation of the X-24B crew, DeMar lost his X-24A crew chief, Jim Hankins, to the new Digital-Fly-

By-Wire (DFBW) program that, using an F-8 fighter as a test-bed, would create the world's first fully digital fly-by-wire aircraft (i.e., one without a mechanical back-up system). In the F-8 fighter used in the DFBW program was a reprogrammed version of the computer used earlier to control the Apollo Lunar Landing Vehicle, another example of the sort of cost-savings practiced at the Flight Research Center by recycling equipment from earlier projects into new ones.

Many of the X-24B crew recruited by Norm DeMar had experience with rocket-powered aircraft, having been on the crews of either the X-24A or the X-15. Charley Russell, a crew chief on the X-15, became crew chief for the X-24B. Three of the X-24A mechanics—Mel Cox, John Gordon, and Ray Kellogg—were assigned as well to the X-24B crew. Other X-24B crew members included inspector Bill Bastow, instrumentation inspector Dick Blair, and aircraft inspector Gaston Moore.

DeMar and the X-24B crew managed to install systems in the X-24B and prepare for systems tests by February 1973, just three and a half months after Martin Marietta had delivered the X-24B as an empty shell. Rather than full-scale wind-tunnel tests, a very detailed set of eleven types of ground and captive-flight tests was scheduled by the two X-24B program managers, NASA's Jack Kolf and the Air Force's Johnny Armstrong, to be done during the six months between February and August 1973 before the first glide flight.

During structural resonance tests on the X-24B's control system, we found an unacceptable resonance in the ailerons. It was a purely mechanical resonance, sustained solely by the actuator and its linkage. To eliminate it, we added a mechanical damper to the actuator's servo valve.

We ran ground vibration tests on the horizontal and vertical tail surfaces to verify flutter clearance margins. Since the results were significantly different from the predicted mathematical model used by Martin Marietta, we reran the flutter analysis using the experimentally determined model data, finding flutter margins to be adequate.

To establish the relationships between applied loads and strain gauge responses, we did structural loads calibration tests on all ten movable control surfaces as well as on the left fin and strake. For use later in interpreting flight results, we also measured the outputs of strain gauges and derived the appropriate load equations.

As had been done on the earlier lifting bodies, the X-24B was hung at different angles to determine the vehicle's center of gravity, then crosschecked by weighing the vehicle while it was balanced on each wheel and tipped at various angles. We used the "rocking table" technique to determine pitch and roll inertias. The vehicle was also hung on a cable and oscillated, using springs attached at both ends of the vehicle, to determine yaw inertia and the product of inertia, the coupling between roll and yaw.

On the X-24B, we expected very high landing-gear loads during X-15-like "slap-down" landings due to its long nose, forward center-of-gravity relative to the location of the main gear, and its increased weight—1,800 pounds more than the X-24A. To provide additional tire capability, we had selected 12-ply T-38 tires for the X-24B, rather than the 10-ply tires used on the X-24A. During dynamic load tests on the tires at Wright Patterson Air Force Base, however, the tread repeatedly separated from the tire casing at the anticipated loading. Later tests showed that shaving the tread from the tire through the first ply resulted in satisfactory tire performance. As a result, we decided that a new set of shaved tires would be used for each flight of the X-24B.

During drag-load tests on the main gear, we found that the down-load lock released when predicted drag loads were applied, which could result in gear collapse during landing. The crew reworked the locking device so that it would maintain a securely locked position.

We did "slap-down" tests on the nose gear to verify the strength of the new back-up structure as well as the energy-absorbing capability of the nose gear and new metering pins in the X-24B. For these drop tests, we elevated the nose of the vehicle with the main tires restrained and then released the vehicle from increasing heights. To produce appropriate nose-gear drag loads, we rotated the nose tires with a spin-up device prior to each release. During these tests, the structure and nose-gear performance proved to be satisfactory.

Flutter while the X-24B hung in pre-launch position under the wing of the B-52 could cause structural failure on the B-52. Therefore, vibration tests were conducted on the B-52 with the X-24B hanging in launch position that assured us that no flutter would occur in flight from the B-52's wing, the lifting-body adapter, or the X-15 pylon.

We ran a series of taxi tests with incremental increases in speed to test for nose-gear shimmy, which we felt was possible due to the X-24B's nose-gear steering that made it distinctly different from the other lifting bodies. The other lifting bodies had had nonsteerable dual nose wheels that avoided all possible shimmy problems. Our primary concern with the X-24B's nose-gear steering was that the nose gear or backup structure might fail if severe shimmy occurred in the nose gear at touchdown on the first flight, given the dynamic load added to the already high landing loads that we expected.

Eight taxi runs were made at speeds from 40 to 150 knots, using the main LR-11 rocket engines as well as the 500-pound hydrogen-peroxide rockets intended to help the pilot during the landing flare. The 150-knot run across the lakebed runway was made using approximately 4,000 pounds of thrust from two LR-11 chambers. Even at 150 knots, the nose-gear steering and ground handling characteristics of the X-24B were found to be satisfactory, with no shimmy in the nose gear. However, lateral center-of-gravity was offset two inches during the test run, the liquid-oxygen tank on the left side outweighing by 1,000 pounds the alcohol fuel tank on the right, making the X-24B pull to the left. The pilot was able to compensate for the offset with intermediate right braking.

We made a final taxi test to 80 knots on the take-off runway with the X-24B hanging under the B-52. Both accelerometer measurements and comments from the pilot verified that the ride was smooth and no problems could be predicted.

During the captive-flight test of the X-24B, we had to exercise much greater care than we had in captive-flight tests of the other lifting bodies, for the pilot of the X-24B could not eject while the vehicle was mated to the B-52. To obtain acceptable loads on the forward hook of the X-15 pylon, we located the X-24B adapter further aft under the pylon than we had with the other lifting bodies, a design compromise based on the proven safe operation of the X-24A.

If there had been a problem during the captive flight, X-24B pilot John Manke would have had to launch before he could have ejected safely. The B-52 was flown as slowly as possible during the climb to 30,000 feet, where structural resonance tests were conducted at speeds higher than possible on the ground. Since the X-24B was within gliding distance of the dry lakebed during these tests, Manke could have landed the vehicle if it had broken loose or had had to be launched, but no problems occurred during the single captive flight.

Flight Tests of the X-24B

On 1 August 1973, John Manke piloted the X-24B on its first glide flight, launching from the B-52 at 40,000 feet, coasting earthward at 460 miles per hour, performing a series of maneuvers to establish handling qualities, and executing a practice landing flare approach before making a 200-mile-per-hour landing on the lakebed. On the flight, the same flight-test maneuvers and evaluations were done that had been done on flights of the earlier lifting bodies. During the series of glide flights that followed, Manke and Major Michael V. Love, the Air Force X-24B project pilot, checked the vehicle's performance in a variety of configurations.

On 15 November 1973, John Manke piloted the X-24B in its first powered flight. Typical flight time in the X-24B was seven minutes, longer than in the other lifting bodies. As pilots had done before flights in the earlier lifting bodies, Manke and Love completed pre-flight practices of numerous simulated approaches in the T-38 and F-104 aircraft. By the end of the X-24B project, lifting-body pilots had flown more than 8,000 such simulated approaches in support of the entire lifting-body program.

On 25 October 1974, during the sixteenth flight of the X-24B, Love reached the aircraft's fastest flight speed, Mach 1.75—or 1,164 miles per hour. On 22 May 1975, Manke made the X-24B's highest approach and landing, coming down to the lakebed from 74,100 feet—more than 14 miles above the earth's surface.

Love and Manke were pleasantly surprised by the handling qualities of the X-24B at all speed ranges, both with and without engaging the control dampers in the stability augmentation system. Even in turbulence, the X-24B flew surprisingly well. In subsonic handling qualities, the X-24B earned the very high rating of 2.5 on the Cooper-Harper pilot rating scale. In short, the X-24B was considered a fine aircraft.

Manke and Love said the handling characteristics of the X-24B compared favorably with those of the fighter aircraft, the T-38 and F-104. The X-24B's handling and riding qualities in turbulence during the final approach were superior to those of the earlier lifting bodies. The high dihedral effect of the other lifting bodies had created disconcerting roll upsets for pilots due to sideslips in turbulence. With its low values in effective dihedral, the X-24B rode turbulence with more of a side-to-side motion

X-24B simulating future Shuttle landings. The F-104 chase plane is behind and to the (pilot's) right of the X-24B. (NASA photo EC75 4914)

that the pilots found more acceptable. The pilots also found the vehicle's dampers-off handling qualities in the landing pattern to be excellent, commenting that they could not believe the dampers were off.[4]

Despite the fact that the X-24B was 1,800 pounds heavier than the X-24A, it had achieved a top speed of Mach 1.75 due to the lower configuration drag of the X-24B and a 15 percent increase in thrust from the uprated LR-11 engines. Although the X-24A had reached Mach 1.6, it very likely could have achieved Mach 1.7 had its test-flight program not been cut short to build the X-24B.

X-24B Simulations of Future Shuttle Landings

By the time that the Space Shuttle was well into the design phase, space mission planners wanted to know if such unpowered re-entry shapes with low lift-to-drag ratios could land successfully on asphalt or concrete runways. Convinced that the X-24B could successfully execute such an approach and landing, John Manke had recommended even earlier that the X-24B make a series of landings on Runway 04/22, the main 15,000-foot concrete runway at Edwards. For John Manke, Mike

4. See John A. Manke and M. V. Love, "X-24B Flight Test Program," *Society of Experimental Test Pilots, Technical Review* 13 (Sept. 1975): 129-54.

Love, and other pilots, such a demonstration seemed important for developing the confidence needed to proceed with similar landings of the Space Shuttle.

In January 1974, the X-24B research subcommittee had approved Manke's proposal. Afterwards, Manke and Love began a three-week flight program, flying the F-104 and T-38 in landing approaches approximating those of the X-24B. Manke alone made over 100 of these approaches.

The payoff came on 5 August 1975. Manke launched in the X-24B from the B-52 mothership, climbed to 60,000 feet, began his descent, and—seven minutes after launch—touched down in the X-24B precisely at the planned target landing spot, 5,000 feet down Runway 04/22. Afterwards, Manke said, "We now know that concrete runway landings are operationally feasible and that touchdown accuracies of ± [plus or minus] 500 feet can be expected."[5] Assisting landing accuracy, Manke commented, were the distance markers and geographic features along the concrete runway, not characteristic at the time of the lakebed runways. Two weeks after Manke's first runway landing, Love duplicated the feat in the X-24B.

These precise touchdowns demonstrated to the Shuttle program that a configuration with a comparatively low lift-to-drag ratio could land accurately without power, thereby convincing Shuttle authorities that they could dispense with the airbreathing jet engines originally planned for the Orbiters. The resultant reduction in weight added significantly to the Shuttle's payload.

Of all the vehicles flight-tested during the twelve years of the lifting-body program, the X-24B had the highest landing lift-to-drag ratio, 4.5. Next highest was the X-24A at 4.0, then the HL-10 at 3.6. Lowest among the lifting bodies was the M2-F3 with a landing lift-to-drag ratio of 3.1. Because of its relatively high lift-to-drag ratio plus good control characteristics, the X-24B was considered by the pilots to be very comfortable to land without power. The lifting-body pilots also considered the M2-F3 acceptable in landing characteristics, although the M2-F3 required more of the pilot's attention in landing, due to having less time from the flare to setting the wheels down on the runway.

By the end of the X-24B program, we had gained widespread experience with the unpowered landing characteristics of lifting-body configurations over a range of landing aerodynamic performance. In its maximum "dirty" configuration—with flaps, deployed landing gear, speed brakes, and low levels of thrust—the F-104 had been used to train pilots in landing approaches for both the X-15 and lifting-body programs, beginning in 1959 with NASA pilots Neil Armstrong (of Apollo fame) and Joe Walker. During the course of these F-104 flights, the aircraft would be landed in the worst lift-to-drag configuration—with flaps, gear, and speed brakes extended in idle power—that approached a maximum lift-to-drag ratio of 2.5. Later tests conducted by Bob Hoey and the Air Force pilots concluded that landing without aids, a vehicle with a

5. *Ibid.*, p. 140.

maximum lift-to-drag ratio of 2.5 bordered on the totally unacceptable—that is, a landing where the risk of crashing is highest. These test results in the F-104 served as a benchmark for the Flight Research Center while evaluating the different flight-tested lifting-body configurations for future space operations.

Landing performance and safety were critical as well in terms of the ablative heat shields used for re-entry vehicles before the development of such new heat-protection materials as the lightweight silicon tiles. We tailored the concept of the lifting body as a re-entry vehicle to the use of the ablative heat shields, the technology current at the time. As a result, landing performance and safety were linked to how the roughness resulting from the burned and melted ablative heat shields would affect the aerodynamic drag of the lifting bodies.

We had excellent data from flight tests at hypersonic speeds made during the X-15 program to use in predicting the magnitude of this effect for the lifting bodies, available in Lawrence C. Montoya's ***Drag Characteristics Obtained from Several Configurations of the Modified X-15-2 Airplane up to Mach 6.7***.[6] The report compares the drag characteristics of a clean-surfaced X-15 with an X-15 flown with an ablative coating. We also had the results of a similar test done on the X-24A during the full-scale wind-tunnel testing of the vehicle at NASA Ames. Although the X-24A was later flight-tested at the Flight Research Center only with a clean metal skin, the wind-tunnel testing of the X-24A with a coating simulating the ablative roughness typical after the heat of re-entry showed a significant reduction for the vehicle in landing lift-to-drag ratio. This reduction, in turn, would reduce significantly the time a pilot would have for making corrections in control during an actual landing of a lifting-body re-entry vehicle.

When we used the ablative roughness data from the X-15 and the X-24A tests to calculate the aerodynamics of lifting bodies with ablative roughness, we found that some lifting-body configurations previously found to be acceptable for flight would become unacceptable as re-entry vehicles with ablative roughness. The ablative roughness after the heat of re-entry would cause the drag of lifting bodies to increase between 15 and 30 percent, lowering the lift-to-drag performance. As a result, for example, the 3.1 lift-to-drag ratio of the M2-F3 would be lowered to less than 2.5, making the M2-F3 unacceptable as a re-entry vehicle unless considerable care were taken to use the correct heat-protection materials in certain places, such as carbon-carbon rather than ablative material on the leading edges. Likewise, the HL-10's lift-to-drag ratio of 3.4 would drop to a ratio that would make it barely acceptable in re-entry. With ablative roughness added, the only lifting bodies that would retain adequate lift-to-drag ratios would be the X-24A and X-24B.

When Bill Dana made the last powered flight of the X-24B on 23 September 1975, the lifting-body program drew to a close. After Dana's flight, six pilot familiarization

6. Lawrence C. Montoya, ***Drag Characteristics Obtained from Several Configurations of the Modified X-15-2 Airplane up to Mach 6.7*** (Washington, DC: NASA TM X-2056, 1970).

glide flights were made in the X-24B by Air Force Captain Francis R. Scobee and NASA's Einar Enevoldson and Tom McMurtry. On 26 November 1975, piloted by McMurtry, the X-24B completed its 36th and final flight. Through the spring of 1976, before being sent to the Air Force Museum, the X-24B remained in residence at Edwards Air Force Base, resplendent in its blue and white paint scheme.

CHAPTER 9

WINGLESS FLIGHT LIVES ON

When Tom McMurtry landed the X-24B for the last time in November 1975, NASA's lifting-body program officially ended. Yet the legacy of wingless flight has lived on, continuing to have a significant impact on the design and technology of current and developing vehicles. In the 1980s and 1990s, the lifting-body legacy went international as Russia, Japan, and France began to design and test lifting bodies. During the early 1990s, the USA began to develop lifting-body designs for use as space-station transports, as spacecraft, and as a future replacement of the current Space Shuttle.

Today, meanwhile, the original lifting-body vehicles flight-tested at the NASA Flight Research Center in the 1960s and early 1970s are all in museums, in outdoor mounted displays, or in the process of being restored for future public displays.

The first lifting body—the M2-F1, fondly dubbed "the flying bathtub"—was stored outdoors at NASA Dryden for several years. It was damaged when it was blown over by the wind, but it was in the process of being restored as these lines were written. Several of the craftsmen who built the M2-F1 have contributed their time and labor in restoring it to its exact original condition. Eventually, the M2-F1 may conceivably join the M2-F3—the third lifting body, built from the crashed M2-F2—at the Smithsonian National Air and Space Museum in Washington, D.C., where the M2-F3 now hangs from the ceiling.

The HL-10 is currently on display outdoors at NASA Dryden, mounted atop a pedestal. Earlier, the HL-10 had been severely damaged when it was dropped by a crane that was lifting it off a truck for display at the Los Angeles Museum of Science. The nose and vertical tails were smashed when the HL-10 hit the ground. Fortunately, no one was hurt in the accident. However, those of us who had worked on the lifting-body program were understandably upset with the Museum's crane crew and handlers, given the HL-10's illustrious flight-test record of 37 flights without mishap or damage. Jerry Reedy and his expert team in NASA Dryden's "Skunk Works" sheet-metal shop repaired the HL-10, doing the work in their spare time. Restored to its original condition, the HL-10 was carefully and sturdily mounted on its current pedestal display at NASA Dryden, never again to be lifted by a crane.

For several years, the Hyper III was stored outdoors at NASA Dryden, along with the M2-F1. Fully restored by the Air Force, the Hyper III now hangs from the ceiling in the Air Museum at Castle Air Force Base near Merced, California.

The X-24B is in the Air Force Museum at Wright Patterson Air Force Base in Dayton, Ohio. The original X-24A was converted into the X-24B, but to show what the X-24A looked like, a mocked-up SV-5J configured as the X-24A is displayed next to the X-24B. ASSET and PRIME, recovered following successful re-entries, are also displayed near the X-24B.

M2-F1 being disassembled for restoration in 1994. People from viewer's left to right: Bill Dana, Dale Reed, Dan Garrabrant, Dick Fischer, and Dick Klein, all of whom participated in the original lifting-body program. (NASA photo EC94 42484-2)

Lifting-Body Concept Goes International

The NASA lifting-body program has been well documented in about 100 technical reports on the program's 222 flights and 20,000 hours of wind-tunnel tests. Many of these publications are unclassified. The Soviet Union purchased copies of these reports from NASA Headquarters in Washington, D.C., then designed its own lifting body. In 1982, the Soviets flight-tested an unpiloted, 10-foot-long, subscale version of their lifting body, the BOR-4, including a maneuvering re-entry over the Indian Ocean from space orbit. The flight test of the BOR-4 closely resembled that of our PRIME (X-23) vehicle in 1966. The main difference between the two was that the BOR-4 was parachuted into the Indian Ocean for ship recovery, while the PRIME had been snatched from the air by a C-130 to avoid a splash down in the Pacific Ocean. There is no information available yet on whether the successor states of the former Soviet Union continued their work in the 1980s into larger-scale lifting bodies.

Other than the Soviet flight experiments with the BOR-4, very little lifting-body design activity seems to have occurred in the 1980s. In the United States, the Shuttle satisfied all requirements for space flight. Even though the Soviets had built a copy of our Shuttle, they flew it only once (unpiloted) and continued to rely on parachute recovery for their spacecraft.

As the concept of an International Space Station emerged in the United States and other countries, however, interest revived in lifting-body configurations. Countries involved in or entering space exploration accepted that the International Space Station was required if mankind were to maintain a presence in space in preparing to send human beings to Mars and other planets or to revisit the moon. Smaller vehicles costing less to operate would be needed over many decades for the International Space Station. The small, compact shapes of lifting-body configurations began to show up on engineering sketch pads and drawing boards for use as space ferries or emergency lifeboats.

Other countries entering the realm of space technology have also demonstrated interest in lifting bodies for various projected space missions. For example, the Japanese conducted hypersonic re-entry flight tests with parachute recovery during February 1997 of the HYFLEX, an unpiloted lifting body. The French also have indicated that they are doing mission studies using lifting bodies.

HL-20 Lifting-Body Space Ferry

During 1990–1995, NASA Langley conducted studies on a new lifting-body shape designated the HL-20, designed to meet the projected need for a low-cost transportation system to ferry personnel between Earth and future space stations. As part of the NASA Langley study, personnel at North Carolina Agricultural and Technical University and North Carolina State University built a full-scale mockup of the HL-20 in 1990.

Designed to carry ten people into orbit and back, the HL-20 would be considerably smaller and lighter than the present Shuttle, the large size of the Shuttle being unnecessary for many of these future missions involved with transporting personnel to and from space stations and with delivering supplies to space stations. The projected HL-20 would be only 31 feet long and weigh 32,448 pounds, considerably smaller and lighter than the Shuttle, which is about 122 feet long and weighs over 171,000 pounds without its propellants, external tank, and solid rocket boosters. The HL-20 would be boosted into orbit by a Titan III rocket system, a system smaller than the Shuttle's rocket system.

National AeroSpace Plane, the X-30 Lifting Body

During the first half of the 1990s, while NASA Langley was conducting studies on the HL-20, several government organizations including NASA were conducting substantial studies on the design of a National AeroSpace Plane (NASP/X-30) capable of taking off from an aircraft runway, flying into space, and returning to Earth for a landing on an airport runway—all without the use of rocket boosters. Of the various configurations studied, a form of lifting body emerged that integrated a hypersonic air-breathing propulsion system within the vehicle's shape. However, the NASP study was terminated in 1994, when it was concluded that the high-temperature materials and air-breathing propulsion technology required for such prolonged high speeds within Earth's atmosphere would take many more years to mature than had originally been estimated.

Nevertheless, NASA continued efforts on its own toward developing a space transportation system that would eventually replace the Shuttle. Opting to stay with rocket-propulsion systems, NASA required a design that would emphasize maximum efficiency, combining a very efficient single-stage-to-orbit vehicle with an advanced rocket-engine system built into the vehicle's shape.

From X-33 to VentureStar

In 1994, NASA solicited proposals from the aerospace industry for designing and building a highly efficient single-stage-to-orbit vehicle to replace the Shuttle. Three proposals were submitted, one each from McDonnell Douglas, Rockwell AeroSpace, and Lockheed Martin.

McDonnell Douglas submitted a design for a lifting body that took off vertically and landed vertically on its tail. The design was reminiscent of Disneyland's Future Space Ride, where millions of tourists have sat in a simulator cabin watching a screen showing the spacecraft lifting off a launch pad and returning to the pad on landing, a procedure very similar to that shown in the Buck Rogers movies of the past. McDonnell Douglas built and flew a scale model of its proposed DC-X rocket to prove that it could indeed take off and land vertically on its tail.

Rockwell AeroSpace proposed a design that was a conservative but highly refined remake of the current Shuttle. The proposed vehicle had wings, could be propelled into space vertically without drop-off boosters, and could land horizontally.

By contrast, the design proposed by Lockheed Martin pushed the state-of-the-art in rocket propulsion, integrating a rocket motor design within the shape of a lifting-body spacecraft. NASA decided that this proposal reduced risk more than the other two alternatives. As a result, NASA awarded a contract valued at about $1 billion from 1996 through 2000 to the Lockheed Martin Skunk Works in Palmdale, California, for building and flight-testing the X-33, a 67-foot-long prototype model of the projected 127-foot-long VentureStar. The X-33's empty weight will be about one-ninth that of VentureStar. The 2,186,000-pound lift-off weight of the proposed VentureStar is about one-half of the 4,500,000-pound lift-off weight of the Shuttle.

According to David Urie, then manager for high-speed programs at the Lockheed Martin Skunk Works, his firm had put together a design that would launch vertically like a rocket and land horizontally like an airplane. Lockheed Martin had decided against a horizontal take-off because the weight of the undercarriage would limit the

X-33 Advanced Technology Demonstrator designed to demonstrate in flight the new technologies needed for a Reusable Launch Vehicle for the 21st century. The X-33 employs a lifting-body shape. (NASA photo EC96 43631-2)

payload severely. Vertical landing was rejected because the vehicle would have to carry fuel for landing throughout flight.[1]

Wings were also eliminated during the design process. Although a lifting body is not as efficient as a winged aircraft at producing lift, the advantage of a lifting-body design for VentureStar is that—minus fuel and payload—the vehicle will be very light when landing back on Earth, light enough to land on an 8,000-foot-long runway, which is shorter than those found at most major airports today. Another significant feature of the lifting body is the large amount of storage area.

The X-33 prototype as well as VentureStar will use the "aerospike" rocket engine developed by the Boeing Company's Rocketdyne Division of Canoga Park, California. Whereas conventional rocket engines have round bell-shaped nozzles, the aerospike engine uses the changing ambient air pressure as the rocket ascends to regulate the

1. See, e.g., Bill Sweetman, "Venture Star: 21st Century Space Shuttle," ***Popular Science*** (October 1996): 44; "Reusable Launch Vehicle," in ***Spinoff 1996*** (Washington, D.C.: NASA, [1996]): 30–31 for printed accounts of the X-33 and Venture Star plans, which were still in the developmental stages as this account was written. David Urie read this section for the author and expressed his satisfaction with its accuracy. In addition, Stephen D. Ishmael, Deputy Manager for X-33 Flight Test and Operations, representing NASA at the Lockheed Martin Skunk Works in Palmdale, read the section and offered some changes that have been incorporated in the narrative.

shape of the exhaust plume. The conventional nozzle, on the other hand, operates at its highest level of efficiency only at a single altitude.

In the 1960s, Rocketdyne developed a rounded aerospike nozzle, leading to the "linear aerospike" engine in 1972, with the gas stream exiting along the surface of a rectangular wedge rather than around a round spike-shape. Its designers hoped that NASA would use the new engine to power the Shuttle, but NASA opted at that time for a more conservative design. The engine fairly much sat on the shelf until Urie and his colleagues at Lockheed Martin mated it with a lifting-body design in creating the X-33 and VentureStar's initial designs.

Seven of the linear aerospike engines will be arrayed across the entire trailing edge of VentureStar, the engines blending into the lifting-body shape. According to its designers, this arrangement will cause less drag in descent than that caused by a cluster of conventional engines. As with most rockets, VentureStar will be steered during ascent by vectoring engine thrust. However, unlike conventional rockets that move on gimbals, VentureStar's engines are fixed, the rockets' automatic flight-control system adjusting the throttle on each engine's upper and lower modules to steer the vehicle.

Beneath the rocket's carbon-fiber skin, tanks on each side carry liquid hydrogen. A smaller tank in the nose contains liquid oxygen, which mixes with the hydrogen for combustion. Located in the middle of the vehicle is a 45-by-15-foot payload bay.

The vehicle has been designed to lift 40,000 pounds of payload to low Earth orbit and 25,000 pounds to the higher orbit occupied by a space station, most of the liftoff weight consisting of the liquid hydrogen and oxygen propellants. With the airframe, engines, and flight-control systems making up only nine percent of the proposed vehicle's 2.2 million-pound liftoff weight, science writer Bill Sweetman has said VentureStar is "roughly equivalent to a 20-pound racing bike carrying a 200-pound rider."[2]

Launching VentureStar should be dramatically different from today's space launches of the Shuttle, with considerably savings in time and materials as well as increased safety. VentureStar will not use the solid rocket boosters that, with the current Shuttle, must be fished out of the ocean and rebuilt after each Shuttle flight. Furthermore, VentureStar will use a metal heat shield, eliminating the 17,000 hours of between-flights maintenance currently involved in checking and replacing heat-resistant ceramic tiles on the outer surface of the Shuttle.

Because of the large surface area of the lifting body, its designers expect VentureStar to re-enter Earth's atmosphere more gently than does the current Shuttle. Unlike the Shuttle's maneuvering thrusters, which use hypergolic propellants that ignite on contact with one another, VentureStar will use only liquid hydrogen and oxygen for propellants. Unlike the Shuttle, VentureStar will have no hydraulic system, using electrically powered flight controls, doors, and landing gears.

---

2. Sweetman, "Venture Star," p. 46.

Currently, Shuttle launches can be delayed when NASA engineers discover a glitch in a satellite payload already loaded onto the launcher. VentureStar bypasses this potential problem by using a self-contained canister as a payload bay. The satellite's manufacturer will load it into the canister, test it, and deliver it to the NASA launch site.

The currently cumbersome job of assembling the Shuttle vehicle on a vertical tower is eliminated with VentureStar, for the vehicle will have no boosters or external tanks. This will allow VentureStar to be checked out in a hangar, like an airplane. Furthermore, VentureStar is expected to be safer than today's rockets, its design reducing the potential for catastrophic problems. While conventional rockets are doomed if an engine fails in flight, VentureStar's engines have a thrust reserve for emergency use. If one of VentureStar's seven engines should fail on liftoff, the engine opposite it would shut down to balance the spacecraft, the remaining five engines then throttling up to carry the vehicle safely into orbit.

Employing a reusable rocket in the design of VentureStar is not only safer but friendlier to the environment. Its exhaust is comprised of water vapor, not the chemical wastes produced by a solid rocket, and there are no spent boosters to create a trail of debris behind the rocket.

Flight-testing the X-33 prototype is expected to resolve certain critical issues before Lockheed Martin begins to build the full-scale VentureStar. For instance, although the X-33 will not fly to orbital speed, it will fly fast enough to test the aerodynamics and metal heat shield under realistic conditions.

As in the X-15 and lifting-body programs between 1959 and 1975, NASA Dryden will play a major role in flight-testing the X-33. The plan is to begin flight-testing the X-33 at Edwards Air Force Base (AFB) in 1999. The X-33 is expected to reach Mach 3 on its first flight before landing at one of the small dry lakebeds northeast of Edwards AFB. Fifteen flight tests are planned at speeds up to Mach 15, mostly between Edwards AFB in Southern California and Malmstrom AFB near Great Falls, Montana.

Designing and building VentureStar is expected to begin in 2000, and the partners in the venture hope it will fly in 2004 as a commercial vehicle. The second VentureStar, ready for flight by 2006, might be the first to carry astronauts. If VentureStar proves it can fly as often as is currently projected, possibly only three or four vehicles would need to be built. Once VentureStar is fully operational, there likely will be a number of VentureStar launch and recovery sites around the world, each site considerably smaller than today's launch sites.[3]

3. Sweetman, "Venture Star," p. 47; "Reusable Launch Vehicle," *Spinoff 1996* , p. 31.

Space-Station Rescue Vehicle

In 1992, I met up again with my old friend John Kiker at an annual international parachute conference in San Diego, California. Years ago at the NASA Johnson Space Center, Kiker had been responsible for developing the parachute systems used in the Gemini and Apollo programs. He had long since retired from NASA but was working part-time as a consultant to NASA on the design of parachute systems for spacecraft.

Kiker introduced me at the conference to Rob Meyerson, the young engineer who at that time headed parachute research and development at the Johnson Space Center. Over lunch, Meyerson told me that there was interest at the Johnson Space Center in developing a lifeboat that would remain attached to the International Space Station for use in case of a need for emergency evacuation.

Ideally, Meyerson said, the lifeboat would be totally automatic in flight, from de-orbit through re-entry and landing. Something more efficient than the Russian Soyuz two-to-three-person re-entry vehicle, recovered with a symmetrical parachute, was desirable. The ideal space-station lifeboat for Meyerson and his colleagues at the Johnson Space Center would use a guidance system allowing personnel to quickly punch landing coordinates into the lifeboat's onboard computer before or after boarding the vehicle. After the lifeboat had separated from the space station, onboard computers would fire the retro rockets at the right time during orbit for landing at the designated site on Earth.

Dale Reed pictured with the X-38 technology demonstrator for a crew return vehicle from the International Space Station and the subscale model used in a test program for the X-38. (NASA photo EC97 44152-5)

A lifting-body design would allow the lifeboat to fly during re-entry to a landing site 700 to 800 miles left or right of the orbital path. After the vehicle had decelerated to subsonic speed at about 20,000-feet altitude, a series of parachutes would be deployed—symmetrical deceleration chutes followed by a large, rectangular-shaped, gliding parafoil parachute. With a gliding ratio of about 3.5 to 1, the parafoil parachute could be steered left and right by two lines attached to winches inside the vehicle. Global Positioning System (GPS) satellites would provide navigation to the landing site.

At an altitude of about 1,000 feet, the onboard computer would command a landing pattern with a downwind leg, base, and final approach into the wind. Using a sonar, radar, or laser altitude ground sensor, the computer would then command both winches to reel in the trailing edges of the parafoil parachute. Next, a landing flare maneuver would reduce the parachute sink rate from about twenty feet per second to less than five feet per second. With a parachute loading of about two pounds per square foot, the no-wind gliding speed would be about 40 miles per hour, slowing at touchdown to less than 30 miles per hour. Such low speed at landing would allow the vehicle to touch down off-runway, such as on any flat field free of obstacles.

After Meyerson had related this information to me, I mentioned that I had been involved in some model flight tests of this concept in 1969 at the NASA Flight Research Center, except we had used the limp Rogallo Parawing gliding parachute with lifting-body shapes rather than the parafoil gliding parachute (see Chapter 8). Fascinated by the idea of GPS guidance, I told Meyerson that I would like to prove the concept by air-launching a lifting-body model as I had in 1969. After Meyerson returned to the Johnson Space Center, he had $150,000 sent to NASA Dryden so that I could put together a team to demonstrate the recovery concept using a subscale model.

During the spring of 1992, we began the test program. By the end of summer, using a team of four and working part-time, we had achieved fully autonomous flight, including flared landing into the wind at a predetermined landing site in the Mojave Desert. Alex Sim served as NASA research project engineer. I did the design work and flight-planning. Jim Murray handled the electronics and data analysis. David Neufeld not only did the parachute rigging and packing but also served as radio-control pilot when the autonomous guiding system was disengaged during developmental flight-testing.

Neufeld became so enthusiastic about his role as pilot that he took sky-diving lessons to learn more about controlling parafoil parachutes. He made only two static-line jumps during his sky-diving lessons, but both were stand-up landings in the center of the 600-foot circle used as a landing zone. I asked him why he hadn't made more jumps. He told me that he had learned from the two jumps all that he needed to learn about piloting the model, so why push his luck?

To study the feasibility of the system, we used a flight model of a spacecraft in the generic shape of a flattened biconic (an object shaped like two cones with their bases together). The model weighted about 150 pounds and was flown under a commercial ram-air parachute. Key elements of the system included GPS navigational guidance, flight-control computer, ultrasonic sensing for terminal altitude, electronic compass, and onboard data-recording.

The vehicle was developed and refined during the flight-test program. It completed autonomous flight from an altitude of 10,000 feet and a lateral offset of 1.7 miles, ending with a precision flare and landing into the wind at the predetermined site. At times during autonomous flight, wind speed nearly equaled vehicle airspeed. We also evaluated several novel techniques for computing winds postflight. In September 1993, we published the results of these tests in NASA Technical Memorandum 4525, ***The Development and Flight Test of a Deployable Precision Landing System for Spacecraft Recovery***.[4]

This was the first time I had worked with a fully autonomous air vehicle. I found myself talking to it as if I were coaching an onboard student pilot. As the model reached a planned turning point in the sky, I would say to it, "Now turn! Now turn!" As it approached for landing, I found myself telling it, "Now flare! Now flare!"

Meyerson discussed the results of our model tests with John Muratore, an engineering colleague at the Johnson Space Center. Muratore had recently become famous for organizing a "pirate team" that developed a low-cost spacecraft control room by using personal computers. His control room had just been pressed into service to operate the Shuttle in flight, saving NASA millions of dollars through fewer controllers and substantially lower maintenance costs on computer and display systems.

Muratore became very interested in the lifeboat concept and presented it to NASA Headquarters, enhancing his proposal by selecting a tried-and-proven lifting-body shape—that of the X-24A—for the lifeboat development program. The X-24A was the only lifting body that had been proven in flight from near-orbital speeds to horizontal landing. Although the unpiloted X-23 PRIME had demonstrated maneuvering flight from orbital speeds down to Mach 2, it was the X-24A that had then demonstrated flight from Mach 2 to subsonic landing speeds.

His selection of the X-24A lifting-body shape also saved on costs, avoiding the need to develop a new spacecraft shape. NASA Headquarters bought the idea that Muratore would prove the concept in low-cost steps to help in making management decisions for later steps leading to launching a prototype into space.

Muratore telephoned me to see what I thought about the proposal and stipulation, especially whether I thought NASA Dryden would be willing to support the Johnson Space Center in a low-cost, full-scale flight demonstration of the lifting-body parafoil-parachute-recovery concept. I said that during the lifting-body program, the NASA

---

4. Alex G. Sim, James E. Murray, David Neufeld, and R. Dale Reed, ***The Development and Flight Test of a Deployable Precision Landing System for Spacecraft Recovery*** (Washington, D.C.: NASA TM 4525, 1993). Both John Muratore, NASA's project manager for the X-38 at the Johnson Space Center and William H. (Bill) Dana, Dryden's chief engineer and former lifting-body pilot, read this chapter, as did Gray Creech, Dryden aerospace projects writer. The narrative has been improved in several places by their comments.

Flight Research Center had spent twelve years proving lifting bodies in horizontal landing. Consequently, NASA Dryden tended to be biased in favor of landing lifting bodies horizontally on runways rather than using a gliding-parachute landing. Later, during a telephone conference among Muratore, NASA Dryden director Ken Szalai, and Szalai's management staff, this bias became apparent, especially with Bill Dana, one of the world's most experienced lifting-body pilots, now serving as Szalai's chief engineer.

Muratore explained to Szalai and the others that studies at the Johnson Space Center had clearly shown that the lifeboat concept utilizing parachute recovery was the most effective in cost and time for rescuing astronauts from the International Space Station. During the studies, Muratore's team had considered several different basic schemes, including a capsule and a horizontal-landing mini-shuttle. With a capsule, to land at an acceptable site, astronauts might have to wait as long as 18 hours in orbit, substantially increasing life-support requirements for the vehicle. With a mini-shuttle, on the other hand, the tail would lose control authority, "blanked" by the high angle of attack during re-entry into Earth's atmosphere, requiring complicated maneuvering engines.

Muratore also explained the added costs involved with both the ocean recovery of parachuted capsules and the horizontal landing of high-speed lifting bodies. The first involves the high cost of maintaining ocean ships to rescue the capsules. The second involves the maintenance of long runway landing facilities.

X-38 suspended under the pylon that would attach it to the B-52 mothership for later captive flights and launches. Note that the X-38 has an X-24A shape. (NASA photo EC97 44105-29)

X-38 suspended under B-52 0008 on its first captive flight, July 30, 1997. (NASA photo EC97 4416316)

To keep lifting-body landing speeds low, Muratore explained that the vehicles would have to be either lighter or larger in size for the same weight. However, the larger lifting bodies would not be compatible with current rocket launch systems, such as the Ariane 5, Titan 4, and possibly the Atlas 2AS, Delta 3, H-2, Proton D-1, or Zenit as well. The 24-foot-long X-24A, for instance, had usually landed after fuel exhaustion at a weight near 6,000 pounds, although Bill Dana said he once made an emergency landing in the M2-F3 with a gross weight of 10,000 pounds.

To be compatible with boosters, Muratore said, the lifting-body spacecraft recovery vehicles must be kept small but weigh 16,000 pounds or more due to internal systems and payloads. A lifting body with such high density would normally require extremely high horizontal landing speeds, too high to be acceptable to Muratore's lifeboat designers. However, the use of a large parafoil gliding parachute could reduce landing speeds to a very low 40 miles per hour, opening up the potential for off-runway landings around the world.

Szalai's team agreed to commit NASA Dryden to helping Muratore and the Johnson Space Center with the program. Szalai asked how Dryden could help. Muratore asked that it furnish and operate its B-52 for launching the Johnson Space Center's experimental vehicle at Edwards AFB. Szalai agreed.

According to the agreement, Dryden would design and build a wing pylon so its B-52 could carry the experimental vehicle aloft. Besides operating the B-52, Dryden would also furnish its ground and hangar facilities and be responsible for personnel and range safety. Johnson Space Center, on the other hand, would be responsible for designing and fabricating the experimental vehicle or vehicles. In this way, a new lifting-body flight-test program—the X-38—came to NASA Dryden, its first in nearly twenty years.

For building three full-scale fiberglass models of the X-38 lifting body, the Johnson Space Center contracted with Scaled Composites, Inc., Burt Rutan's little airplane factory in Mojave, California, not far from Edwards AFB. The three vehicles included one without fins for launching from a C-130 plus two with fins and control surfaces for launching from the B-52.

In the spring of 1995, with the assistance of the Army, the first vehicle was launched from a C-130 over the parachute-testing range at Yuma, Arizona. An extraction chute pulled the finless lifting-body on an aluminum cargo pallet rearward from the C-130. Immediately after launch from the C-130, a problem developed with the cargo pallet and the parachute rigging. The pull from the extraction chute deformed the cargo pallet, causing parachute rigging deflections. Out-of-sequence line cutter and parachute deployments followed. The parachute system became entangled, and the first X-38 vehicle was destroyed on ground contact.

Scaled Composites, Inc., completed the other two X-38s in the fall of 1996, delivering them to the Johnson Space Center for systems installation. Flight tests began at NASA Dryden in the summer of 1997.

By the end of 1997, it is hoped that a successful flight demonstration can be made—launching an X-38 from the B-52 at 45,000 feet, the X-38 then flying as a lifting body in controlled flight down to 20,000 feet, where a series of pilot chutes, drag chutes, and the large 7,300-square-foot parafoil gliding parachute will deploy. The X-38 would then be steered and flared autonomously to a landing site on the dry lakebed at Edwards AFB.

Following successful flight demonstrations from B-52 launches, Muratore plans a follow-on vehicle built of aluminum with a shell of graphite-cyanate ester epoxy. Improved and larger Shuttle-derived blankets and tiles will provide thermal protection to the vehicle's stiffer composite structure. The plan is to launch this vehicle into orbit in 2000 from the Space shuttle. After this vehicle is successfully recovered from orbit, the plan is to build four to eight mission vehicles designed to carry astronauts and service the International Space Station.

A Lifetime of Excitement and Adventure

Little did I know in 1962—as I was flying those paper models of lifting bodies in the hallways at NASA Dryden and later the first balsa models on a ranch east of Lancaster in California—that I would see major flight-test and spacecraft lifting-body programs come into being within the decade. Still less did I know then that, as these programs came into being, I would get to know and have the opportunity to work with the greatest minds and human spirits in aerospace—from designers of airplanes and spacecraft to the best pilots, flight crews, and technicians in the world.

Our work during the 1960s and early 1970s in developing and flight-testing the first experimental lifting bodies has had a highly significant influence on decisions guiding the course of events in the space program. For instance, the decision to develop the Shuttle as an unpowered glider was heavily influenced by our flight experience at the NASA Flight Research Center with the lifting bodies. Because lifting-body landings had proved that unpowered landings were not only safe but reliable, the Shuttle design did not include the extra weight of deployable turbojet engines necessary for powered landings. The reduced weight increasing the Shuttle's carrying capacity significantly.

There are now immediate and direct applications on the horizon for lifting-body vehicles. Although a lifting-body configuration has not yet emerged as an operational vehicle, that reality is getting very close and is now within sight. Wingless flight—both in and out of Earth's atmosphere—is now a firm and substantiated technology, thanks to the hard work and dedication of the men and women involved with the lifting-body concept during the 1950s, 1960s, and 1970s. Most of us who were involved at that time are today retired or nearing retirement, passing the legacy of wingless flight on to the next generation of engineers, scientists, technicians, pilots, and astronauts. Our legacy exists in detail for this new generation, recorded in numerous technical reports and flight-test records. The young engineers of today, who will carry flight innovation into the 21st century, can make solid and informed decisions in considering a wingless configuration for future space systems.

In writing this book, I wished to give the new generation something that isn't always obvious when reading technical reports and flight-test records. I wanted them to know that those reports and records were produced by real people with very human feelings who shed much sweat, some tears, and even some blood in arriving at the facts and data that might seem coldly detached from human realities on the printed page. When I recall the very high risks we sometimes took during the twelve years of initial lifting-body history, I know for certain that we could have spilled much more blood than we did. I prefer to think that even as we were pushing things to the edge, we were smart enough not to fall off and needed only a little luck to protect us from ourselves.

APPENDIX

Lifting Body Flight Log

Part One: Light Weight, M2-F1

Light Weight Lifting Body Flight Log (M2-F1)

Date	GRD Tow	Air Tow	Pilot	Free Flight Sec	Tow Vehicle	Tow Vehicle Pilot	Remarks
3/1/63	2		Thompson	None	PONTIAC		First Ground Tow
4/5/63	11		Thompson	None	PONTIAC		First Airborne Time
4/23/63	10		Thompson	0:00:13	PONTIAC		First Free Flight
8/16/63		1	Thompson	0:02:00	R4D	Mallick/Dana	First Air Tow
8/28/63		1	Thompson	0:22:09	R4D	Mallick/Dana	
8/29/63		1	Thompson	0:02:25	R4D	Mallick/Dana	
8/30/63		2	Thompson	0:04:42	R4D	Mallick/Dana	
9/3/63		2	Thompson	0:04:50	R4D	Mallick/Dana	
10/7/63		1	Thompson	0:01:26	R4D	Butchart/Dana	
10/9/63		1	Thompson	0:01:51	R4D	Haise/McKay	
10/15/63		1	Thompson	0:02:20	R4D	Butchart/?	
10/23/63		1	Thompson	0:03:00	R4D	Butchart/McKay	
10/25/63		2	Thompson	0:03:52	R4D	Butchart/Mallick	
11/8/63		3	Thompson	0:07:45	R4D	Mallick/McKay/ Butchart	
12/3/63		1	Thompson	0:01:00	R4D	Dana/Mallick	
12/3/63		1	Yeager	0:01:35	R4D	Dana/Mallick	
12/3/63		2	Peterson	0:03:15	R4D	Dana/Mallick	Broke Main Wheels
1/29/64		2	Thompson		R4D	Dana/McKay	
1/29/64		2	Peterson	0:04:44	R4D	Dana/McKay	
1/29/64		2	Yeager		R4D	Dana/McKay	
1/30/64		2	Yeager		R4D	Dana/McKay	

APPENDIX

1/30/64	2	Mallick		R4D	Dana/McKay	
2/28/64	2	Thompson		R4D	Butchart/Peterson	
3/30/64	1	Peterson	0:02:25	R4D	Butchart/Kluever	Fired Landing Rocket
4/9/64	2	Thompson		R4D	Butchart/Kluever	
4/9/64	3	Peterson	0:08:00	R4D	Butchart/Kluever	
5/19/64	2	Peterson	0:04:08	R4D	Butchart/McKay	Rocket Landing Asst.
6/3/64	1	Thompson		R4D	Dana/Peterson	
7/24/64	3	Peterson	0:06:50	R4D	Dana/Haise	2- Flts, Rockets Used
8/18/64	1	Thompson		R4D	Dana/Peterson	
8/21/64	4	Thompson		R4D	Dana/Haise/Walker	
2/16/65	1	Thompson		R4D	Dana/Peterson	Airspeed Calib.
5/27/65	4	Thompson		R4D	Butchart/Haise	
5/27/65	3	Sorlie	0:06:00	R4D	Butchart/Peterson	
5/28/65	1	Thompson		R4D	Haise/Peterson	
5/28/65	2	Sorlie	0:04:30	R4D	Peterson/Haise	
7/16/65	1	Thompson		R4D	Haise/Kluever	
7/16/65	1	Dana		R4D	Haise/Kluever	
7/16/65	1	Gentry	0:00:09	R4D	Haise/Kluever	1st Slow Roll
8/30/65	3	Thompson		R4D	Peterson/Haise	
8/31/65	1	Thompson		R4D	Haise/Peterson	
10/6/65	2	Thompson		R4D	Peterson/Haise	
10/8/65	1	Thompson		R4D	Haise/Peterson	
3/28/66	2	Thompson		R4D	Peterson/Butchart	
8/4/66	1	Peterson	0:02:00	R4D	Butchart/Fulton	
8/5/66	3	Peterson	0:04:00	R4D	Butchart/Fulton	
8/10/66		Gentry		PONTIAC		Final Car Tow
8/16/66	1	Gentry		R4D	Butchart/Fulton	2nd Slow Roll
8/18/66	Project Cancelled by Paul Bikle					

Note: There were approximately 400 tows by the Pontaic, but not all of them were recorded.

Compiled by Betty Love, converted to Pagemaker format by Dennis DaCruz

Part Two: Heavy Weights, M2-F2, M2-F3, HL-10, X-24A, X-24B

Veh.	Date	Pilot	Mach Number	Miles Per Hour*	Altitude	Remarks
M2-F2	7/12/66	Thompson	0.646	452	45,000	First Lifting Body (L/B) Free-flight
M2-F2	7/19/66	Thompson	0.598	394	45,000	
M2-F2	8/12/66	Thompson	0.619	430	45,000	
M2-F2	8/24/66	Thompson	0.676	446	45,000	
M2-F2	9/2/66	Thompson	0.707	466	45,000	Thompson's last L/B flight, 360 degree approach
M2-F2	9/16/66	Peterson	0.705	466	45,000	Peterson's 1st L/B flight
M2-F2	9/20/66	Sorlie	0.635	421	45,000	Sorlie's 1st L/B flight
M2-F2	9/22/66	Peterson	0.661	436	45,000	
M2-F2	9/28/66	Sorlie	0.672	443	45,000	
M2-F2	10/5/66	Sorlie	0.615	430	45,000	Sorlie's last L/B flight
M2-F2	10/12/66	Gentry	0.662	436	45,000	Gentry's 1st L/B flight
M2-F2	10/26/66	Gentry	0.605	399	45,000	
M2-F2	11/14/66	Gentry	0.681	445	45,000	
M2-F2	11/21/66	Gentry	0.695	457	45,000	
HL-10	12/22/66	Peterson	0.693	457	45,000	Limit Cycle/Flow Separation; Unmodified HL-10
M2-F2	5/2/67	Gentry	0.623	411	45,000	
M2-F2	5/10/67	Peterson	0.612	403	45,000	Peterson's last L/B Flight; Landing Accident
HL-10	3/15/68	Gentry	0.609	425	45,000	Mod II-Gentry's 1st HL-10 flight
HL-10	4/3/68	Gentry	0.690	455	45,000	
HL-10	4/25/68	Gentry	0.697	459	45,000	
HL-10	5/3/68	Gentry	0.688	454	45,000	
HL-10	5/16/68	Gentry	0.678	447	45,000	

HL-10	5/28/68	Manke	0.657	434	45,000	Manke's 1st L/B flight
HL-10	6/11/68	Manke	0.635	433	45,000	
HL-10	6/21/68	Gentry	0.637	435	45,000	
HL-10	9/24/68	Gentry	0.682	449	45,000	XLR-11 Engine Installed
HL-10	10/3/68	Manke	0.714	471	45,000	
HL-10	10/23/68	Gentry	0.666	449	39,700	1st Powered Flt., Eng. Malf., Landed Rosamond
HL-10	11/13/68	Manke	0.840	524	42,650	3 Tries to Light Engine
HL-10	12/9/68	Gentry	0.870	542	47,420	
X-24A	4/17/69	Gentry	0.718	474	45,000	Glide Flight
HL-10	4/17/69	Manke	0.994	605	52,740	
HL-10	4/25/69	Dana	0.701	462	45,000	Dana's 1st L/B Flight
X-24A	5/8/69	Gentry	0.693	457	45,000	Glide Flight
HL-10	5/9/69	Manke	1.127	744	53,300	1st Supersonic L/B Flight
HL-10	5/20/69	Dana	0.904	596	49,100	
HL-10	5/28/69	Manke	1.236	815	62,200	
HL-10	6/6/69	Hoag	0.665	452	45,000	Hoag's 1st L/B Flight
HL-10	6/19/69	Manke	1.398	922	64,100	
HL-10	7/23/69	Dana	1.271	839	63,800	
HL-10	8/6/69	Manke	1.540	1020	76,100	1st 4-chambered Flight
X-24A	8/21/69	Gentry	0.718	486	40,000	Glide Flight
HL-10	9/3/69	Dana	1.446	958	77,960	
X-24A	9/9/69	Gentry	0.594	402	40,000	
HL-10	9/18/69	Manke	1.256	833	79,190	
X-24A	9/24/69	Gentry	0.596	396	40,000	Glide Flight
HL-10	9/30/69	Hoag	0.924	609	53,750	
X-24A	10/22/69	Manke	0.587	387	40,000	Manke's 1st X-24 Flight

HL-10	10/27/69	Dana	1.577	1,041	60,610	
HL-10	11/3/69	Hoag	1.396	921	64,120	
X-24A	11/13/69	Gentry	0.646	427	45,000	Glide Flight
HL-10	11/17/69	Dana	1.594	1,052	64,590	
HL-10	11/21/69	Hoag	1.432	952	79,280	
X-24A	11/25/69	Gentry	0.685	454	45,000	Glide Flight
HL-10	12/12/69	Dana	1.310	871	79,960	
HL-10	1/19/70	Hoag	1.310	869	86,660	
HL-10	1/26/70	Dana	1.351	897	87,684	
HL-10	2/18/70	Hoag	1.861	1,228	67,310	Fastest L/B flight
X-24A	2/24/70	Gentry	0.771	509	47,000	
HL-10	2/27/70	Dana	1.314	870	90,303	Highest L/B flight
X-24A	3/19/70	Gentry	0.865	571	44,400	1st Powered X-24 Flight
X-24A	4/2/70	Manke	0.866	571	58,700	
X-24A	4/22/70	Gentry	0.925	610	57,700	
X-24A	5/14/70	Manke	0.748	494	44,600	Only 2 Chambers Lit
M2-F3	6/2/70	Dana	0.688	469	45,000	1st M2-F3 Flight
HL-10	6/11/70	Hoag	0.744	503	45,000	Lift/Drag Powered Approach
X-24A	6/17/70	Manke	0.990	653	61,000	
HL-10	7/17/70	Hoag	0.733	499	45,000	Hoag's/HL-10's Last Flight
M2-F3	7/21/70	Dana	0.660	440	45,000	
X-24A	7/28/70	Gentry	0.938	619	58,100	
X-24A	8/11/70	Manke	0.986	651	63,900	
X-24A	8/26/70	Gentry	0.694	458	41,500	Only 2 Chambers Lit
X-24A	10/14/70	Manke	1.186	784	67,900	1st Supersonic X-24 Flight
X-24A	10/27/70	Manke	1.357	899	71,400	Highest X-24 Flight
M2-F3	11/2/70	Dana	0.630	429	45,000	
X-24A	11/20/70	Gentry	1.370	905	67,600	

M2-F3	11/25/70	Dana	0.809	534	51,900	1st M2 -F3 Powered Flight
X-24A	1/21/71	Manke	1.030	679	57,900	
X-24A	2/4/71	Powell	0.659	435	45,000	Powell's 1st L/B Flight
M2-F3	2/9/71	Gentry	0.707	469	45,000	Gentry's 1st M2-F3/Last L/B Flights
X-24A	2/18/71	Manke	1.511	998	67,400	
M2-F3	2/26/71	Dana	0.773	510	45,000	Only 2 Chambers Lit
X-24A	3/8/71	Powell	1.002	661	56,900	
X-24A	3/29/71	Manke	1.600	1,036	70,500	Fastest X-24 Flight
X-24A	5/12/71	Powell	1.389	918	70,900	Delayed Light of Rocket Chamber
X-24A	5/25/71	Manke	1.191	786	65,300	Only 3 Chambers Lit
X-24A	6/4/71	Manke	0.817	539	54,400	Only 2 Chambers Lit/Last X-24 A Flight
M2-F3	7/23/71	Dana	0.930	614	60,500	
M2-F3	8/9/71	Dana	0.974	643	62,000	
M2-F3	8/25/71	Dana	1.095	723	67,300	1st Supersonic M2-F3 Flight
M2-F3	9/24/71	Dana	0.728	480	42,000	Engine Malfunction, Fire, Rosamond Landing
M2-F3	11/15/71	Dana	0.739	487	45,000	New Jettison Location Checkout
M2-F3	12/1/71	Dana	1.274	843	70,800	
M2-F3	12/16/71	Dana	0.811	535	46,800	Only 2 Chambers Lit
M2-F3	7/25/72	Dana	0.989	652	60,900	1st Command Augmentation System Flight
M2-F3	8/11/72	Dana	1.101	726	67,200	
M2-F3	8/24/72	Dana	1.266	835	66,700	
M2-F3	9/12/72	Dana	0.880	581	46,000	Engine Malfunction, Small Fire
M2-F3	9/27/72	Dana	1.340	885	66,700	
M2-F3	10/5/72	Dana	1.370	904	66,300	100th Lifting Body Flight

M2-F3	10/19/72	Manke	0.905	597	47,100	Manke's 1st M2-F3 Flight
M2-F3	11/1/72	Manke	1.213	803	71,300	
M2-F3	11/9/72	Powell	0.906	597	46,800	Powell's 1st M2-F3 Flight
M2-F3	11/21/72	Manke	1.435	947	66,700	Planned Rosamond Lakebed Landing
M2-F3	11/29/72	Powell	1.348	890	67,500	
M2-F3	12/6/72	Powell	1.191	786	68,300	Planned Rosamond Lakebed Landing
M2-F3	12/13/72	Dana	1.613	1,064	66,700	Fastest M2/Used L/D Rockets
M2-F3	12/20/72	Manke	1.294	856	71,500	Highest and last M2-F3 Flight
X-24B	8/1/73	Manke	0.640	460	40,000	First Glide Flight of X-24B
X-24B	8/17/73	Manke	0.650	449	45,000	
X-24B	8/31/73	Manke	0.716	495	45,000	
X-24B	9/18/73	Manke	0.687	450	45,000	
X-24B	10/4/73	Love	0.704	461	45,000	Love's 1st L/B Flight
X-24B	11/15/73	Manke	0.930	598	52,764	1st X-24B Powered Flight
X-24B	12/12/73	Manke	0.987	645	63,081	
X-24B	2/15/74	Love	0.696	456	45,000	
X-24B	3/5/74	Manke	1.086	708	60,334	1st X-24B Supersonic Flight
X-24B	4/30/74	Love	0.876	578	52,040	Love's 1st Powered Flight
X-24B	5/24/74	Manke	1.140	753	55,979	
X-24B	6/14/74	Love	1.228	810	65,512	
X-24B	6/28/74	Manke	1.391	920	68,150	
X-24B	8/8/74	Love	1.541	1,022	73,380	
X-24B	8/29/74	Manke	1.098	727	72,440	
X-24B	10/25/74	Love	1.752	1,164	72,150	Max. Speed/X-24B Flight
X-24B	11/15/74	Manke	1.615	1,070	72,060	
X-24B	12/17/74	Love	1.585	1,036	68,780	
X-24B	1/14/75	Manke	1.748	1,157	72,787	

X-24B	3/20/75	Love	1.443	955	70,373	
X-24B	4/18/75	Manke	1.204	795	57,900	
X-24B	5/6/75	Love	1.444	958	73,400	
X-24B	5/22/75	Manke	1.633	1,084	74,100	
X-24B	6/6/75	Love	1.677	1,110	72,100	
X-24B	6/25/75	Manke	1.343	887	58,000	
X-24B	7/15/75	Love	1.585	1,047	69,480	
X-24B	8/5/75	Manke	1.190	773	60,000	1st Runway Landing./Manke's Last Flight
X-24B	8/20/75	Love	1.548	1,010	72,000	Runway Landing/Love's Last Flight
X-24B	9/9/75	Dana	1.481	990	71,000	
X-24B	9/23/75	Dana	1.157	780	58,000	Last Rocket Flight/Dana's Last Flight
X-24B	10/9/75	Enevoldson	0.705	450	45,000	Enevoldson's 1st L/B Flight
X-24B	10/21/75	Scobee	0.696	462	45,000	Scobee's 1st L/B Flight
X-24B	11/3/75	McMurtry	0.702	456	45,000	McMurtry's 1st L/B Flight
X-24B	11/12/75	Enevoldson	0.702	456	45,000	Enevoldson's Last L/B Flight
X-24B	11/19/75	Scobee	0.700	460	45,000	Scobee's Last L/B Flight
X-24B	11/26/75	McMurtry	0.713	460	45,000	McMurtry's Last L/B Flight/Last X-24B Flight

* Approximate

Assembled from a compilation by Jack Kolf and Appendix N of Richard P. Hallion, ***On the Frontier: Flight Research at Dryden, 1946-1981*** (Washinton, D.C.: NASA SP-4303, 1984); formatted in Pagemaker by Dennis DaCruz.

GLOSSARY

ablation	Thermal process where the surface melts or vaporizes at high temperature, thereby absorbing heat created aerodynamically.
ablator	Surface material that will melt or vaporize to absorb heat.
active cooling	Process whereby a heat-conductive fluid circulates between a hot and cool region, drawing off heat.
ADP	Advanced Development Projects—a Lockheed group located in California.
AF or USAF	United States Air Force.
AFB	Air Force Base.
AFFTC	Air Force Flight Test Center.
AFSC	Air Force Systems Command, an Air Force major command during the period of this narrative.
analog computer	In the context of this book, a computer in a simulator that solves equations of motion using analogous electrical circuits; that is, it expresses data in terms of measurable quantities, such as voltages, rather than by numbers as a digital computer does.
AOA	Angle of Attack: direction of relative wind with respect to an aircraft's longitudinal axis.
Apollo	NASA program to land a human on the moon and return to earth.
ARC	NASA Ames Research Center.
ASD	Aeronautical Systems Division (Air Force).
aspect ratio	The ratio of squared airfoil length (span) to total airfoil area or of airfoil length to its mean chord (distance from leading to trailing edge). Thus, an airfoil of high aspect ratio is relatively long with a relatively short chord, whereas one of low aspect ratio is comparatively short and stubby.

GLOSSARY

attitude The position or orientation of an aircraft or spacecraft with relation to its axes and some reference line or plane.

ballistic Adjective describing the path of a body launched into a trajectory where it is subject only to the forces of gravity and drag.

ballistic coefficient Weight divided by the drag coefficient times the frontal area.

bank angle Angle between the plane of an aircraft's wings and the horizon

boat-tail Shape of the rear of a vehicle whose cross section decreases from the center to the aft end.

C-130 Four-engine, turboprop-powered transport airplane.

capsule A self-contained, symmetrical container capable of safely entering the earth's atmosphere from orbital or higher speeds.

CD Drag coefficient. A non-dimensional parameter for measuring drag.

c.g. Center of gravity—an imaginary location within an object that identifies its center of mass.

ceramic tiles Small blocks of rigid material (primarily silica) attached to the outside of a gliding re-entry vehicle that prevent the heat generated by re-entry speeds from reaching the vehicle structure.

chase planes Aircraft used to fly close to research airplanes for purposes of providing the research pilot with an additional set of eyes for safety purposes.

chord The straight-line distance from the leading to the trailing edge of an airfoil such as a wing.

CL Lift coefficient. A non-dimensional parameter for measuring lift.

CLS/W Lift coefficient divided by wing loading. A non-dimensional parameter that allows the glide performance of several aircraft to be compared at the same airspeed.

control laws — The relationship between the pilot's commands and the actual control surface (aileron, elevon, etc.) movements produced by a flight control system.

cross range — The distance that can be achieved by a re-entry vehicle (as it enters the atmosphere) in a direction perpendicular to that of the initial entry path.

damp — To slow down.

decouple mode — An entry concept that uses a different deceleration method for entry than for landing.

delta wing — A wing that has a triangular shape when viewed from above.

digital — Adjective describing a mechanism, such as a computer, that expresses data in discrete, numerical digits.

dihedral — Effect on lifting bodies of sideslip, producing roll.

DoD — Department of Defense.

doublet — An aircraft control movement from neutral to a deflected position that is held, then returned in the opposite direction back to the original neutral position.

drag — A force that resists motion and is produced by friction within the atmosphere.

Dutch roll — A complex oscillating motion of an aircraft involving rolling, yawing, and sideslipping—so-named from the resemblance to the characteristic rhythm of an ice skater.

Dyna-Soar — Short for Dynamic Soaring. Name of a boost-glide research program that was canceled in 1963 before its first flight. The aircraft designation was X-20A.

effective dihedral — An aircraft aerodynamic characteristic that makes the airplane roll (rotate around the longitudinal axis) when a sideslip or side gust is encountered.

eyeballs-in — A descriptive term used to identify the direction of a force due to acceleration.

F-104	Air Force century series fighter built by Lockheed and used as a chase and research airplane at the Flight Research Center for many years.
FDL	The Air Force Flight Dynamics Laboratory located at the Wright-Patterson Air Force Base in Dayton, Ohio.
FDL-7	Seventh re-entry design created at the FDL.
FDL-8	Eighth re-entry design created at the FDL.
fineness ratio	The ratio of body length to body width of an aerodynamic shape.
flight cards	A type of check list in card form used by pilots and other crew members to track events in a planned flight test.
flight path	The path of a moving object, usually measured in the vertical plane relative to the horizon.
fly-by-wire	A flight control concept that uses only electrical signals between the pilot's stick and the control surfaces.
FRC	The NASA Flight Research Center located at Edwards, California. From 1954 to 1959, the designation of this organization was the NACA and then the NASA High Speed Flight Station. In 1976, it became the NASA Hugh L. Dryden Flight Research Center.
frontal area	The area of an object as projected onto a plane perpendicular to the flight direction.
gain	Sensitivity with respect to flight controls or a stability augmentation system.
Hyper III	A light-weight, unpiloted vehicle built by the NASA FRC and patterned after the FDL-7 shape.
hypersonic	Characterized by speeds of Mach 5 or greater.
jack points	Designated points marked on the underside of an aircraft wing to push upward with a hydraulic jack for the purpose of calibrating strain gages inside the wing structure.

LaRC	The NASA Langley Research Center located in Hampton, Virginia.
L/D	Lift-to-drag ratio.
Lift	A force on an object produced by aerodynamic reaction with the atmosphere as the object moves; it acts perpendicularly to the flight direction.
limit cycle	A run-away oscillation of an aircraft control surface that occurs when the sensitivity (gain) of the automatic stabilization system is too high.
lower flap control horn	A small mechanical arm attached to a lifting body lower flap control surface to which an actuator control rod is attached.
LOX	Liquid Oxygen.
Mach number	The ratio of an object's speed to that of sound. An object reaches Mach1 when it flies at the speed of sound; Mach 2 is twice the speed of sound; and so forth.
Mercury	First U.S. manned space capsule program.
MLRV	Manned Lifting Re-entry Vehicle. An early NASA Langley Research Center lifting body design.
moment	A tendency to cause rotation about a point or axis, as of a control surface about its hinge.
MOU	Memorandum of Understanding—usually a simple document with signatures stating the agreed-upon responsibilities between two or more organizations.
MSL	Mean Sea Level.
NACA	National Advisory Committee for Aeronautics.
NASA	National Aeronautics and Space Administration.
neutral longitudinal stability	A flight condition in which an aircraft that is disturbed in pitch continues to rotate away from the initial angle of attack at a constant angular rate without returning.

non-receding charring ablator A type of ablator (see above) that maintains its external dimensions while melting or vaporizing.

nose-wheel rotation The point in an aircraft take-off maneuver at which the pilot commands the aircraft to rotate its nose upwards, increasing lift so as to depart the ground.

notch filter An electronic filter in an aircraft's automatic control system to remove or obstruct unwanted frequencies within a narrow band to prevent them from causing problems with the system.

on-the-street The time when an agency advertises (in a request for proposals) that a new job or contract is planned.

overdrive Slang term used to describe the 15 percent increase in thrust that was available on the X-24B rocket engine as compared with that used on previous lifting bodies.

PILOT PIloted LOw speed Test. Early designation for what became the X-24A program.

PIO Pilot Induced Oscillation—a situation in flight in which a pilot causes an aircraft to oscillate about the intended path of flight by making excessive control inputs.

pitch Angular displacement of a vehicle such as an aircraft about the lateral axis (i.e., nose up or nose down).

plow horse The author's term for chubby lifting bodies that are capable of carrying large payloads but have shorter hypersonic cross ranges than race horses (which see).

Pregnant Guppy A C-97 cargo airplane modified to carry an oversized cargo.

PRIME Precision Recovery Including Maneuvering Entry. Early designation for what became the SV-SD or X-23 program.

projected area The area of an object as projected onto a horizontal plane parallel with the flight direction.

PSTS Propulsion System Test Stand.

race horse	The author's term for streamlined, slender lifting bodies with smaller payload capacity than the plow horses (which see) but with very high hypersonic cross ranges.
radiative	A type of cooling that radiates heat away from a cooling hot surface.
ramjet	A type of jet engine without any mechanical compressor, comprised of a specially shaped, open tube into which the air necessary for combustion is forced and then compressed by the forward motion of the aircraft.
rate limited	Term indicating the maximum angular rate at which an actuator can drive an aircraft control surface.
Real Stuff	Term (derived from Tom Wolfe's ***The Right Stuff***) to describe the qualities of people who create and service aircraft or spacecraft for experimental flights rather than fly them.
retrofire	Short-term rocket ignition with the thrust pointed in the direction of flight so as to reduce the speed of an orbiting object and to initiate entry.
Reynolds number	A nondimensional parameter representing the ratio of momentum forces to viscous forces about a body in fluid flow, as in the atmosphere; named for English scientist Osborne Reynolds (1842–1912); among other applications the ratio is vital to the use of wind tunnels for scale-model testing, as it provides a basis for extrapolating the test data to full-sized test vehicles.
Right Stuff	A term first coined by Tom Wolfe in his book of the same title. It refers to the qualities possessed by pilots and astronauts who fly experimental aircraft or spacecraft.
Rogallo Wing	A wing-like parachute design that enables the parachuting object to move forward as well as descend.
roll	Rotational movement of an aircraft or similar body about its longitudinal axis.
roll reversal	An adverse aircraft design condition in which an aircraft rolls in the opposite direction from that commanded by the pilot or control surfaces.

rotation speed The minimum speed at which a pilot can rotate the aircraft nose upward (lift the nose wheel off the runway) during a take-off roll.

RPV Remotely Piloted Vehicle—a vehicle controlled through radio links by a pilot not physically in the vehicle.

RTD Research and Technology Development—an Air Force Organization.

SAMSO Space and Missile Systems Organization—an Air Force organization, part of AFSC during the period covered by this narrative.

SAS Stability Augmentation System—electronic control components designed to augment the stability of an airplane.

second generation vehicle A vehicle that has benefited from the previous design, development and testing of a similar vehicle.

self-adaptive Adjective describing a flight control concept that samples, then alters, internal electronic signals to compensate for changing flight conditions.

semi-ballistic Adjective describing a state in which an object is subject to small aerodynamic forces in addition to the predominant forces of gravity and inertia.

Shuttle The winged vehicle developed by NASA and its contractors to serve as a Space Transportation System to carry cargo to and from earth orbit.

side-arm controller A two- or three-axis control stick mounted on the side of the cockpit and operated by a pilot's wrist movements.

sideslip A sideways movement of an aircraft away from the initial flight path.

simulator A partial aircraft cockpit connected to an electronic computer; it allows a pilot to replicate to a significant degree the flight of an airplane.

Skunk Works Popular term for a small, highly efficient design and fabrication organization capable of creating innovative prototype aircraft in a short period of time. The Lockheed Advanced Development Projects group was the first organization to use the term "Skunk Works" officially to describe its organization.

span The distance from tip to tip or root to tip of an airfoil such as an airplane's wing.

spiral stability A natural aircraft characteristic that allows the vehicle either to remain in level flight or to return thereto when upset in roll or bank angle.

Sputnik 1 The first man-made object to be placed in earth orbit (by the Soviet Union on 4 October 1957).

strain gage An instrument used to measure the strain or distortion in a member or test specimen (such as an aircraft structural part) that is subjected to a force.

strakes Wing-like appendages at the aft end of an aircraft that provide lift or added stability; also long, flat surfaces attached to the exterior of an aircraft's skin and aligned with the local free-stream conditions.

SV-5 Basic configuration of a re-entry vehicle that led to the SV-5P (X-24A) and SV-SD (PRIME).

SV-5J Jet-powered version of the SV-5 configuration. Two were built but neither was flown.

swashplate A mechanical plate with a universal joint giving it freedom to pivot in any direction about one point. Multiple attachment points for control rods in the plane of the plate allowed flexibility for different controls in the M2-F1 lifting body.

test-bed aircraft A conventional aircraft that has been equipped with some newly designed internal or external components for in-flight testing.

Thor-Delta A two-stage rocket using a Thor 1st stage and a Delta 2nd stage.

triply redundant Adjective describing the concept of using three parallel components to accomplish a single function, with automatic de-selection of any faulty component.

tufts Short segments of yarn or string taped to an aerodynamic surface to allow airflow characteristics to be observed directly or photographed.

volumetric efficiency The ratio of total volume to the surface area of a three-dimensional shape. A sphere has the highest volumetric efficiency of any shape.

wedge angle — The angle of the aft control surfaces relative to the flight direction. Large angles produce shuttlecock-like stability.

wetted skin area — The total exposed surface area of any shape. In an aircraft, this is all skin area exposed to the outside airstream.

wing loading — Vehicle weight divided by the area of the wing.

X-24C — A follow-on proposal to the X-24B to test advanced air-breathing propulsion.

yaw — Motion of an aircraft or similar vehicle about the vertical axis (i.e., nose left or right).

Y-plane — Lateral (left to right) axis of an aircraft or flight vehicle.

BIBLIOGRAPHY

Armstrong, Johnny G. "Flight Planning and Conduct of the X-24A Lifting Body Flight Test Program." AFFTC TD-71-10, August 1972.

Baker, David. ***Spaceflight and Rocketry: A Chronology***. New York: Facts on File, 1996.

Brandt, Jerome C. "XLR-11-RM-13 Rocket Engine Development and Qualification Program." AFFTC TD-69-1, August 1969.

Chandler, John. "After Years, Inventor May Get His Shot at Mars." ***Los Angeles Times*** (Dec. 4, 1989): B-3-4.

"Dale Reed: From Models to Mars." Unsigned article in the PRC Inc. company newspaper, May 1990.

Dryden, Hugh L. "Introductory Remarks." National Advisory Committee for Aeronautics, ***Research-Airplane-Committee Report on Conference on the Progress of the X-15 Project***. Papers Presented at Langley Aeronautical Laboratory, Oct. 25-26, 1956.

Grosser, M. "Building the Gossamer Albatross." ***Technology Review***, 83 (Apr. 1981): 52–63.

Hallion, Richard P. ***The Hypersonic Revolution: Eight Case Studies in the History of Hypersonic Technology***. 2 vols.; Wright-Patterson Air Force Base, Ohio: Special Staff Office, 1987.

_____, ***On the Frontier: Flight Research at Dryden***, 1946-1981. Washington, D.C.: NASA SP-4303, 1984.

_____, ***Test Pilots: The Frontiersmen of Flight***. Washington, D.C.: Smithsonian Institution Press, 1981.

Horton, Victor W., Eldredge, Richard C., and Klein, Richard E. "Flight-Determined Low-Speed Lift and Drag Characteristics of the Lightweight M2-F1 Lifting Body." Washington, D.C.: NASA TN D3021, 1965.

Iliff, Kenneth W. "Parameter Estimation for Flight Vehicles." ***Journal of Guidance, Control, and Dynamics*** 12 (Sept.-Oct. 1989): 609-622.

Jex, Henry R. and Mitchell, David G. "Stability and Control of the Gossamer Human-Powered Aircraft by Analysis and Flight Test." Washington, D.C.: NASA Contract Report 3627, 1982.

Kempel, Robert W. "Analysis of a Coupled Roll-Spiral-Mode, Pilot-Induced Oscillation Experienced With the M2-F2 Lifting Body." Washington, D.C.: NASA Technical Note D-6496, 1971.

_____, Dana, William H. and Sim, Alex G. "Flight Evaluation of the M2-F3 Lifting Body Handling Qualities at Mach Numbers from 0.30 to 1.61." Washington, D.C.: NASA Technical Note D-8027, 1975.

_____, Painter, Weneth D., and Thompson, Milton O. "Developing and Flight Testing the HL-10 Lifting Body: A Precursor to the Space Shuttle." Washington, D.C.: NASA Reference Publication 1332, 1994.

Kocivar, Ben. "Superdrone: Pilotless Mite Will Sniff Out Pollution." ***Popular Science*** (July 1978): 66-69.

McCready, Paul, "Crossing the Channel in the Gossamer Albatross." Society of Experimental Test Pilots, ***Technical Review*** 14, No. 4 (1979): 232-43.

Maine, R.E. and Iliff, Kenneth W. "Identification of Dynamic Systems." Washington, D.C.: NASA RP-1138, 1985.

Manke, John A. and Love, M.V. "X-24B Flight Test Program." Society of Experimental Test Pilots, ***Technical Review*** 13 No. 4 (Sept. 1975): 129-154.

Matranga, Gene J. and Armstrong, Neil A. ***Approach and Landing Investigation at Lift-Drag Ratios of 2 to 4 Utilizing a Straight-Wing Fighter Airplane.*** NASA High-Speed Flight Station, Edwards, Calif.: NASA TM X-31, 1959.

Montoya, Lawrence C. ***Drag Characteristics Obtained from Several Configurations of the Modified X-15-2 Airplane up to Mach 6.7.*** Washington, D.C.: NASA TM X-2056, 1970.

National Advisory Committee for Aeronautics. ***NACA Conference on High-Speed Aerodynamics, A Compilation of the Papers Presented.*** Moffett Field, Calif.: Ames Aeronautical Laboratory, 1958.

Pyle, Jon S. and Montoya, Lawrence C. ***Effects of Roughness of Simulated Ablated Material on Low-Speed Performance Characteristics of a Lifting-Body Vehicle***. Washington, D.C.: NASA TM S-1810, 1969.

Reed, R. Dale. "High-Flying Mini-Sniffer RPV: Mars Bound?" ***Astronautics & Aeronautics*** 16 (June 1978): 26-39.

_____. "RPRVs: The First and Future Flights." ***Astronautics & Aeronautics*** 12 (April 1974): 27-42.

"Reusable Launch Vehicle." ***Spinoff 1996***. Washington, D.C.: NASA, 1996. 30-31.

Sim, Alex G., Murray, James E., Neufeld, David, and Reed, R. Dale ***The Development and Flight Test of a Deployable Precision Landing System for Spacecraft Recovery***. Washington, D.C.: NASA TM 4525, 1993.

Sim, Alex G. "Flight-Determined Stability and Control Characteristics of the M2-F3 Lifting Body Vehicle." Washington, D.C.: NASA Technical Note D-7511, 1973.

Smith, Harriett J. "Evaluation of the Lateral-Directional Stability and Control Characteristics of the Lightweight M2-F1 Lifting Body at Low Speeds." Washington, D.C.: NASA Technical Note D-3022, 1965.

Sweetman, Bill. "Venture Star: 21st Century Space Shuttle." ***Popular Science*** (October 1996): 42-47.

Taylor, Lawrence W. and Iliff, Kenneth W. "A Modified Newton-Raphson Method for Determining Stability Derivatives from Flight Data." Paper presented at the Second International Conference on Computing Methods in Optimization Problems, San Remo, Italy, Sept. 9-13, 1968.

Thompson, Milton O., ***At the Edge of Space: The X-15 Flight Program***. Washington, D.C.: Smithsonian Institution Press, 1992.

_____, Peterson, Bruce A., and Gentry, Jerauld R. "Lifting Body Flight Test Program." Society of Experimental Test Pilots, ***Technical Review*** (September 1966): 1–22.

Wallace, Lane E. ***Flights of Discovery: 50 Years at the NASA Dryden Flight Research Center***. Washington, D.C.: NASA SP-4309, 1996.

Wilford, John Noble. "The Playful Inventor: Visions of Future Flight." ***The New York Times*** (Sept. 27, 1983): C1, C6.

Wilkinson, Stephan. "The Legacy of the Lifting Body." ***Air & Space*** (April/May 1991): 50-62.

Winter, Frank. "Black Betsy: The 6000C-4 Rocket Engine, 1945-1989. Part II." ***Acta Astronautica*** 32, No. 4 (1994): 305-308.

Wolfe, Tom, ***The Right Stuff***. New York: Ferrar, Strauss, Giroux, 1979.

ABOUT THE AUTHOR

R. Dale Reed is an aerospace engineer with a bachelor of science degree in mechanical engineering from the University of Idaho. He began working at what later became the NASA Dryden Flight Research Center as a research engineer in 1953. He conducted aerodynamic loads research on the X-1E, X-5, F-100, and D-558-II aircraft and performed aerodynamic heating measurements on the X-15 before he began developing, advocating, and implementing the lifting-body flight research program that he writes about in this book. Before his retirement from NASA in 1985, he won four NASA awards ranging from the Exceptional Service Medal to an Associate Fellow Award.

Following his formal retirement, Mr. Reed went to work for a variety of contractors and is today again working at Dryden, now as a contract engineer. In his long career, he has published 14 technical reports and a number of periodical articles. He has also taken out four patents and managed 19 programs. He is currently working actively with the X-38 program described in these pages.

INDEX

C

D

E

F

G

H

I

J

K

L

M

N

O

P

R

U

V

W

X

Y

Z

THE NASA HISTORY SERIES

Reference Works, NASA SP-4000

Grimwood, James M. ***Project Mercury: A Chronology*** (NASA SP-4001, 1963)

Grimwood, James M., and Hacker, Barton C., with Vorzimmer, Peter J. ***Project Gemini Technology and Operations: A Chronology*** (NASA SP-4002, 1969)

Link, Mae Mills. ***Space Medicine in Project Mercury*** (NASA SP-4003, 1965)

Astronautics and Aeronautics, 1963: Chronology of Science, Technology, and Policy (NASA SP-4004, 1964)

Astronautics and Aeronautics, 1964: Chronology of Science, Technology, and Policy (NASA SP-4005, 1965)

Astronautics and Aeronautics, 1965: Chronology of Science, Technology, and Policy (NASA SP-4006, 1966)

Astronautics and Aeronautics, 1966: Chronology of Science, Technology, and Policy (NASA SP-4007, 1967)

Astronautics and Aeronautics, 1967: Chronology of Science, Technology, and Policy (NASA SP-4008, 1968)

Ertel, Ivan D., and Morse, Mary Louise. ***The Apollo Spacecraft: A Chronology, Volume I, Through November 7, 1962*** (NASA SP-4009, 1969)

Morse, Mary Louise, and Bays, Jean Kernahan. ***The Apollo Spacecraft: A Chronology, Volume II, November 8, 1962–September 30, 1964*** (NASA SP-4009, 1973)

Brooks, Courtney G., and Ertel, Ivan D. ***The Apollo Spacecraft: A Chronology, Volume III, October 1, 1964–January 20, 1966*** (NASA SP-4009, 1973)

Ertel, Ivan D., and Newkirk, Roland W., with Brooks, Courtney G. ***The Apollo Spacecraft: A Chronology, Volume IV, January 21, 1966–July 13, 1974*** (NASA SP-4009, 1978)

Astronautics and Aeronautics, 1968: Chronology of Science, Technology, and Policy (NASA SP-4010, 1969)

Newkirk, Roland W., and Ertel, Ivan D., with Brooks, Courtney G. ***Skylab: A Chronology*** (NASA SP-4011, 1977)

Van Nimmen, Jane, and Bruno, Leonard C., with Rosholt, Robert L. ***NASA Historical Data Book, Vol. I: NASA Resources, 1958–1968*** (NASA SP-4012, 1976, rep. ed. 1988)

Ezell, Linda Neuman. ***NASA Historical Data Book, Vol II: Programs and Projects, 1958–1968*** (NASA SP-4012, 1988)

Ezell, Linda Neuman. ***NASA Historical Data Book, Vol. III: Programs and Projects, 1969–1978*** (NASA SP-4012, 1988)

Astronautics and Aeronautics, 1969: Chronology of Science, Technology, and Policy (NASA SP-4014, 1970)

Astronautics and Aeronautics, 1970: Chronology of Science, Technology, and Policy (NASA SP-4015, 1972)

Astronautics and Aeronautics, 1971: Chronology of Science, Technology, and Policy (NASA SP-4016, 1972)

Astronautics and Aeronautics, 1972: Chronology of Science, Technology, and Policy (NASA SP-4017, 1974)

Astronautics and Aeronautics, 1973: Chronology of Science, Technology, and Policy (NASA SP-4018, 1975)

Astronautics and Aeronautics, 1974: Chronology of Science, Technology, and Policy (NASA SP-4019, 1977)

Astronautics and Aeronautics, 1975: Chronology of Science, Technology, and Policy (NASA SP-4020, 1979)

Astronautics and Aeronautics, 1976: Chronology of Science, Technology, and Policy (NASA SP-4021, 1984)

Astronautics and Aeronautics, 1977: Chronology of Science, Technology, and Policy (NASA SP-4022, 1986)

Astronautics and Aeronautics, 1978: Chronology of Science, Technology, and Policy (NASA SP-4023, 1986)

Astronautics and Aeronautics, 1979–1984: Chronology of Science, Technology, and Policy (NASA SP-4024, 1988)

Astronautics and Aeronautics, 1985: Chronology of Science, Technology, and Policy (NASA SP-4025, 1990)

Gawdiak, Ihor Y. Compiler. *NASA Historical Data Book, Vol. IV: NASA Resources, 1969–1978* (NASA SP-4012, 1994)

Noordung, Hermann. *The Problem of Space Travel: The Rocket Motor.* Ernst Stuhlinger and J.D. Hunley, with Jennifer Garland, editors (NASA SP-4026, 1995)

Management Histories, NASA SP-4100

Rosholt, Robert L. *An Administrative History of NASA, 1958–1963* (NASA SP-4101, 1966)

Levine, Arnold S. *Managing NASA in the Apollo Era* (NASA SP-4102, 1982)

Roland, Alex. *Model Research: The National Advisory Committee for Aeronautics, 1915–1958* (NASA SP-4103, 1985)

Fries, Sylvia D. *NASA Engineers and the Age of Apollo* (NASA SP-4104, 1992)

Glennan, T. Keith. *The Birth of NASA: The Diary of T. Keith Glennan*, edited by J.D. Hunley (NASA SP-4105, 1993)

Seamans, Robert C., Jr. *Aiming at Targets: The Autobiography of Robert C. Seamans, Jr.* (NASA SP-4106, 1996)

Project Histories, NASA SP-4200

Swenson, Loyd S., Jr., Grimwood, James M., and Alexander, Charles C. ***This New Ocean: A History of Project Mercury*** (NASA SP-4201, 1966)

Green, Constance McL., and Lomask, Milton. ***Vanguard: A History*** (NASA SP-4202, 1970; rep. ed. Smithsonian Institution Press, 1971)

Hacker, Barton C., and Grimwood, James M. ***On Shoulders of Titans: A History of Project Gemini*** (NASA SP-4203, 1977)

Benson, Charles D. and Faherty, William Barnaby. ***Moonport: A History of Apollo Launch Facilities and Operations*** (NASA SP-4204, 1978)

Brooks, Courtney G., Grimwood, James M., and Swenson, Loyd S., Jr. ***Chariots for Apollo: A History of Manned Lunar Spacecraft*** (NASA SP-4205, 1979)

Bilstein, Roger E. ***Stages to Saturn: A Technological History of the Apollo/Saturn Launch Vehicles*** (NASA SP-4206, 1980)

Compton, W. David, and Benson, Charles D. ***Living and Working in Space: A History of Skylab*** (NASA SP-4208, 1983)

Ezell, Edward Clinton, and Ezell, Linda Neuman. ***The Partnership: A History of the Apollo-Soyuz Test Project*** (NASA SP-4209, 1978)

Hall, R. Cargill. ***Lunar Impact: A History of Project Ranger*** (NASA SP-4210, 1977)

Newell, Homer E. ***Beyond the Atmosphere: Early Years of Space Science*** (NASA SP-4211, 1980)

Ezell, Edward Clinton, and Ezell, Linda Neuman. ***On Mars: Exploration of the Red Planet, 1958–1978*** (NASA SP-4212, 1984)

Pitts, John A. ***The Human Factor: Biomedicine in the Manned Space Program to 1980*** (NASA SP-4213, 1985)

Compton, W. David. ***Where No Man Has Gone Before: A History of Apollo Lunar Exploration Missions*** (NASA SP-4214, 1989)

Naugle, John E. ***First Among Equals: The Selection of NASA Space Science Experiments*** (NASA SP-4215, 1991)

Wallace, Lane E. ***Airborne Trailblazer: Two Decades with NASA Langley's Boeing 737 Flying Laboratory*** (NASA SP-4216, 1994)

Butrica, Andrew J. Editor. ***Beyond the Ionosphere: Fifty Years of Space Communication*** (NASA SP-4217, 1997)

Butrica, Andrew J. ***To See the Unseen: A History of Planetary Radar Astronomy*** (NASA SP-4218, 1996)

Center Histories, NASA SP-4300

Rosenthal, Alfred. *Venture into Space: Early Years of Goddard Space Flight Center* (NASA SP-4301, 1985)

Hartman, Edwin, P. *Adventures in Research: A History of Ames Research Center, 1940–1965* (NASA SP-4302, 1970)

Hallion, Richard P. *On the Frontier: Flight Research at Dryden, 1946–1981* (NASA SP-4303, 1984)

Muenger, Elizabeth A. *Searching the Horizon: A History of Ames Research Center, 1940–1976* (NASA SP-4304, 1985)

Hansen, James R. *Engineer in Charge: A History of the Langley Aeronautical Laboratory, 1917–1958* (NASA SP-4305, 1987)

Dawson, Virginia P. *Engines and Innovation: Lewis Laboratory and American Propulsion Technology* (NASA SP-4306, 1991)

Dethloff, Henry C. *"Suddenly Tomorrow Came . . .": A History of the Johnson Space Center, 1957–1990* (NASA SP-4307, 1993)

Hansen, James R. *Spaceflight Revolution: NASA Langley Research Center From Sputnik to Apollo* (NASA SP-4308, 1995)

Wallace, Lane E. *Flights of Discovery: 50 Years at the Dryden Flight Research Center* (NASA SP-4309, 1996)

General Histories, NASA SP-4400

Corliss, William R. *NASA Sounding Rockets, 1958–1968: A Historical Summary* (NASA SP-4401, 1971)

Wells, Helen T., Whiteley, Susan H., and Karegeannes, Carrie. *Origins of NASA Names* (NASA SP-4402, 1976)

Anderson, Frank W., Jr. *Orders of Magnitude: A History of NACA and NASA, 1915–1980* (NASA SP-4403, 1981)

Sloop, John L. *Liquid Hydrogen as a Propulsion Fuel, 1945–1959* (NASA SP-4404, 1978)

Roland, Alex. *A Spacefaring People: Perspectives on Early Spaceflight* (NASA SP-4405, 1985)

Bilstein, Roger E. *Orders of Magnitude: A History of the NACA and NASA, 1915–1990* (NASA SP-4406, 1989)

Logsdon, John M., with Lear, Linda J., Warren-Findley, Jannelle, Williamson, Ray A., and Day, Dwayne A., eds. *Exploring the Unknown: Selected Documents in the History of the U.S. Civil Space Program, Volume I: Organizing for Exploration* (NASA SP-4407, 1995)

Logsdon, John M., with Day, Dwayne A., and Launius, Roger D., eds. *Exploring the Unknown: Selected Documents in the History of the U.S. Civil Space Program, Volume II: External Relationships* (NASA SP-4407, 1996)